Human Evolution
AN ILLUSTRATED INTRODUCTION

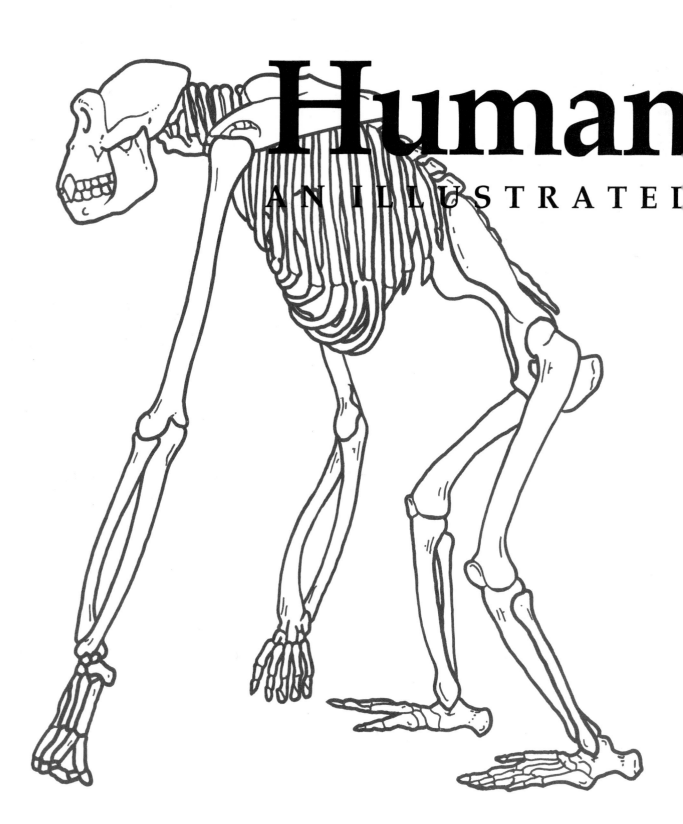

Human

AN ILLUSTRATED

Evolution

INTRODUCTION

ROGER LEWIN

American Association for the
Advancement of Science

SECOND EDITION

BLACKWELL SCIENTIFIC PUBLICATIONS

BOSTON OXFORD LONDON

EDINBURGH MELBOURNE

© 1984, 1989 by
Blackwell Scientific Publications, Inc.
Editorial offices:
3 Cambridge Center, Suite 208
    Cambridge, Massachusetts 02142, USA
Osney Mead, Oxford OX2 0EL, England
8 John Street, London WC1N 2ES, England
23 Ainslie Place, Edinburgh EH3 6AJ, Scotland
107 Barry Street, Carlton
    Victoria 3053, Australia

First published 1984
Reprinted 1985 (twice)
1986, 1987, 1988 (for W.H. Freeman)
Second edition 1989

Set by Setrite Typesetters, Hong Kong
Printed and bound by
The Maple-Vail Book Manufacturing Group
Printed in the United States of America
89 90 91 92 5 4 3 2 1

DISTRIBUTORS

USA and Canada
    Blackwell Scientific Publications, Inc.
    Publishers' Business Services
    PO Box 447
    Brookline Village
    Massachusetts 02147
    (*Orders*: Tel: 617 524–7678)

Canada
    Oxford University Press
    70 Wynford Drive
    Don Mills
    Ontario M3C 1J9
    (*Orders*: Tel: (416) 441–2941)

Australia
    Blackwell Scientific Publications
    (Australia) Pty Ltd
    107 Barry Street
    Carlton, Victoria 3053
    (*Orders*: Tel: 03 347–0300)

Outside North America and Australia
    Blackwell Scientific Publications, Ltd
    Osney Mead
    Oxford OX2 0EL
    England
    (*Orders*: Tel: 011 44 865 240201)

Library of Congress
Cataloging-in-Publication Data

Lewin. Roger.
    Human Evolution: an illustrated introduction/Roger
Lewin.—2nd ed.
        p.   cm.
    Includes Index,
    ISBN 0−86542−066−1.—ISBN 0−86542−067−X
(pbk.)
    1. Human Evolution.   I. Title.
    [DNLM: 1. Evolution.   GN 281 L672h]
GN281.L49   1989
573.2—dc19

Contents

Preface

'Paleoanthropology—the study of human origins—has undergone a transformation in recent years.' This was how, 4 years ago, I began the preface to the first edition of *Human Evolution: an Illustrated Introduction*. Progress in related sciences—in evolutionary biology, behavioral ecology and so on—had combined with a greatly expanded hominid fossil record to give paleoanthropology a sense of scientific security that was absent until not very long ago. The preparation of the second edition, which has involved a virtually complete revision, indicates that the pace of progress has not slowed. What we are seeing now, however, is not so much a transformation, rather a maturation of the science: those new developments that earlier had transformed paleoanthropology are now providing foundations for further insights.

The maturation of many sciences frequently involves an inexorable narrowing of focus, with deeper and deeper issues being tackled at ever increasing detail. With paleoanthropology, however, maturation has demanded the reverse process. No longer is it acceptable—or fruitful—to seek answers to human evolution simply by studying the fossilized remains left behind by our ancestors: obtaining important answers requires a broader perspective.

For instance, we cannot hope to know why brain expansion occurred so dramatically in the human family until we appreciate more clearly the biology of brain expansion within the Primate order in general. And it is now evident that attempting to think about the social organization of our ancestors is futile, until we know more about the interaction between diet and social structure in primates as a whole. In other words, many of the more fundamental questions of human evolution will be answered in the context of primate evolutionary biology and ecology.

One inevitable consequence of this maturing approach to paleoanthropology has been the abandoning of an old equation, specifically, that human origins equals hominid origins. This traditional equation meant that anthropologists looked for humanlike explanations for key events in hominid history, such as in the adoption of bipedalism and the emphasis on brain expansion. These days, humanlike motives and perspectives are seen as rather late entries in hominid prehistory, but exactly how late is still a matter of debate.

Key discoveries during the past 4 years have touched virtually every aspect of hominid biology, including some major fossil finds, developments in evolutionary biology, ecology, and behavior, and in molecular biology and genetics. This latter area—molecular biology and genetics—has already had an important influence in illuminating more clearly the evolutionary relationship among the great apes and humans, and has always promised to extend its impact. That promise is now being fulfilled, with provocative new information about the origin of modern humans. There is surely more to come.

As with the first edition, this revision of *Human Evolution* is intended to provide a broad appreciation of current work and future prospects in paleoanthropology. In addition, where space has allowed, there is an attempt to provide an historical perspective on the questions that anthropologists ask about our origins, and the kinds of answers that have been developed and later rejected. A sense of history is crucial to a recognition that science is a tentative, progressive activity, and what is new and apparently correct now may later turn out to be incomplete or just plain wrong.

It is my privilege to be an observer of this exciting science, and I am indebted to the many practitioners—too numerous to list—who have helped and encouraged me. Here I wish to thank those who kindly reviewed parts of the manuscript and offered advice for improving it: John Fleagle, Robert Foley, Fred Grine, Andrew Hill, William Jungers, Misia Landau, Phyllis Lee, Lawrence Martin, Henry McHenry, Sally McBrearty, Richard Potts, Frank Spencer, and Randall Susman. As always, I am more grateful than words can express to Gail, my wife, for her constant support.

ROGER LEWIN
Washington DC
November 1988

1 / Man's place in nature

The title of this introductory unit derives from a landmark book published in 1863 by Darwin's friend and champion Thomas Henry Huxley: it was called *Evidences as to man's place in nature*. The book, which appeared a little over 3 years after Darwin's *Origin of species*, was based on evidence from comparative anatomy among apes and humans, embryology, and fossil evidence of early humans (of which there was very little available at the time). Huxley's conclusion—that humans share a close evolutionary relationship with the great apes, particularly the African apes—was a key element in a major revolution in the history of western philosophy: humans were to be seen as being *a part of* nature, no longer as *apart from* nature.

Committed though Huxley was to the idea of the evolution of *Homo sapiens* from some type of ancestral ape, he nevertheless considered humans to be a very special kind of animal. 'No one is more strongly convinced than I am of the vastness of the gulf between...man and the brutes,' wrote Huxley, 'for, he alone possesses the marvellous endowment of intelligible and rational speech [and]...stands raised upon it as on a mountain top, far above the level of his humble fellows, and transfigured from his grosser nature by reflecting, here and there, a ray from the infinite source of truth.'

The explanation of this 'gap' between humans and the rest of animate nature has always exercised the minds of Western intellectuals, both in pre- and post-evolutionary eras. One difference between the two eras was that, after Darwin, naturalistic explanations had to account not only for man's physical form but also his exceptional intellectual, spiritual, and moral qualities. Previously, these qualities had been regarded as God-given.

As a result, said the late Glynn Isaac, 'Understanding the literature on human evolution calls for the recognition of special problems that confront scientists who report on this topic', a remark he made at the 1982 centenary celebration of Darwin's death. 'Regardless of how scientists present them, accounts of human origins are read as replacement materials for Genesis. They...do more than cope with curiosity, they have allegorical content, and they convey values, ethics and attitudes.' In other words, in addition to reconstructing phylogenies—or family trees—paleoanthropological research also addresses 'Man's place in nature' in more than just the physical sense. As we shall see, that 'place' has long been regarded as being in some way special.

The revolution wrought by Darwin's work was in fact the second of two such intellectual upheavals within the history of Western philosophy, the first being three centuries earlier when Nicholaus Copernicus replaced the geocentric model of the universe with a heliocentric model. Although the Copernican revolution deposed Man from being the very center of all of God's creation, and instead placed him as the occupant of a small planet cycling in a vast universe, Man nevertheless remained the pinnacle of God's works. And from the sixteenth through to the mid-nineteenth centuries, those who studied Man and nature as a whole were coming close to the wonder of those works.

This pursuit—known as natural philosophy—held science and religion in close harmony, with the remarkable design so clearly manifested in

The Anthropomorpha of Linnaeus: In the mid-eighteenth century, when Linnaeus compiled his *Systema Naturae*, Western scientific knowledge about the apes of Asia and Africa was at best sketchy. Based on tales of sea captains and other transient visitors, fanciful images of these creatures were created. Here, produced from a dissertation of Linnaeus' student Hoppius, are four supposed 'manlike apes', some of which became species of *Homo* in Linnaeus' *Systema Naturae*. From left to right: *Troglodyta bontii*, or *Homo troglodytes*, in Linnaeus; *Lucifer aldrovandii*, or *Homo caudatus*; *Satyrus tulpii*, a chimpanzee; and *Pygmaeus edwardi*, an orangutan.

creatures great and small being seen as evidence of God's hand. In addition to design, a second feature of God's created world was a virtual continuum of form, from the lowest to the highest, with humans being near the very top, just a little lower than the angels. This continuum—known as the Great Chain of Being—was not a statement of dynamic relationships between organisms, reflecting historical connections and evolutionary derivations. Instead, notes Harvard biologist Stephen Jay Gould, 'The chain is a static ordering of unchanging, created entities, a set of creatures placed by God in fixed positions of an ascending hierarchy.'

Powerful though it was, the theory faced some problems, specifically some unexplained gaps. One was that between the world of plants and the world of animals. And another was between humans and apes.

Knowing that the gap between apes and humans should be filled, eighteenth and early nineteenth century scientists tended to exaggerate the humanness of the apes while exaggerating the simianness of some of the 'lower' races. For instance, some apes were 'known' to walk upright, to carry off humans for slaves, and even to produce offspring after mating with humans. By the same token, some humans were 'known' to be brutal savages, equipped with neither culture nor language.

This perception of the natural world inevitably became encompassed within the formal classification system, which was developed by Carolus Linnaeus in the mid-eighteenth century. In his *Systema Naturae*, published first in 1736 and finally in 1758, Linnaeus not only included *Homo sapiens*—the species to which we all belong—but also *Homo troglodytes*, which, though little known, was said to

Ptolemy's Universe: Before the Copernican revolution in the sixteenth century, scholars' view of the universe was based on ideas of Aristotle. The earth was seen as the center of the universe, with the sun, moon, stars, and planets fixed in concentric crystalline spheres circling it.

be active only at night and to speak in hisses, and *Homo caudatus*, about which even less was known, except that it had no tail. 'Linnaeus worked with a theory that anticipated such creatures,' notes Gould, 'since they should exist anyway, imperfect evidence becomes acceptable.' This was not scientific finagling, simply that honest scientists saw what they expected to see, a human weakness that has always operated in science—in all sciences—and always will.

The notion of evolution—the transmutation of species—had been in the air for a long time when, in 1859, the power of data and argument in the *Origin of species* proved decisive. Geological ideas had been changing too, with the notion of Cuvier's 'catastrophism' giving way to the 'uniformitarianism' of Hutton and Lyell. In parallel with this was a revision of the accepted age of the earth, from the 6000 years, implied by calculations from the Bible, to many millions of years, implied by the idea of the slow, steady change of uniformitarianism.

Interestingly enough, although the advent of the evolutionary era brought with it an enormous shift in intellectual perceptions of the *origin* of mankind, many elements concerning the *nature* of mankind remained unassailed. For instance, humans were still regarded as being somehow 'above' other animals, endowed with special qualities—those of intelligence, spirituality, and moral judgment. And the gradation from 'lower' races to 'higher' races that had been part of the Great Chain of Being were now explained by the process of evolution.

'The progress of the different races was unequal', noted Roy Chapman Andrews, a researcher at the American Museum of Natural History in the 1920s and 1930s. 'Some developed into masters of the world at an incredible speed. But the Tasmanians... and the existing Australian aborigines lagged far behind, not much advanced beyond the stages of Neanderthal man.' Such overtly racist comments were echoed frequently in literature of the time, and are to be seen reflected in the published evolutionary trees of the time.

In other words, inequality of races—with blacks on the bottom and whites on the top—was explained away as the natural order of things: before 1859 as the product of God's creation, and after 1859 as the product of natural selection.

In the same vein, there was built into early discussions of human evolution the notion of progress, and specifically the inevitability of *Homo sapiens* as the ultimate aim of it all. 'Much of evol-

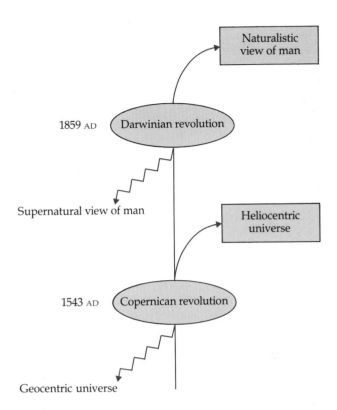

Two great intellectual revolutions: In the mid-sixteenth century the Polish mathematician Nicholas Copernicus proposed a heliocentric rather than a geocentric view of the universe: 'the Earth was not the center of all things celestial,' he said, 'but instead was one of several planets circling a sun, which was one of many sun's in the universe.' Three centuries later, in 1859, Charles Darwin further changed Man's view of himself, arguing that humans were a part of nature, not apart from nature.

ution looks as if it had been planned to result in man, and in other animals and plants to make the world a suitable place for him to dwell in', observed Robert Broom in 1933, who was responsible for some of the more important early human fossil finds in South Africa during the 1930s and 1940s.

Evolution as progress, the steady, inexorable improvement to more complex, more intelligent life: it is, and always has been a seductive notion. 'Progress—or what is the same thing, Evolution—is [Nature's] religion', wrote Britain's Sir Arthur Keith in 1927. In fact, the notion of progress as a driving ethos of nature—and society—has been a characteristic of Western philosophy, but not of all intellectual thought. 'The myth of progress' is how Niles Eldredge and Ian Tattersall characterize it. 'Once evolved, species with their own peculiar adaptations, behaviors, and genetic systems are

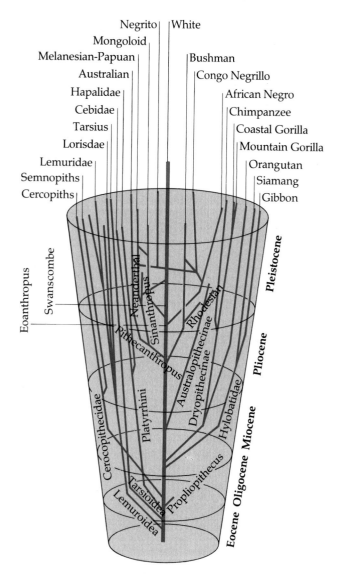

Negrito | White
Mongoloid
Melanesian-Papuan | | Bushman
Australian | | Congo Negrillo
Hapalidae | | African Negro
Cebidae | | Chimpanzee
Tarsius | | Coastal Gorilla
Lorisdae | | Mountain Gorilla
Lemuridae | | Orangutan
Semnopiths | | Siamang
Cercopiths | | Gibbon

Racism in anthropology: In the early decades of this century racism was an implicit part of anthropology, with 'white' races considered to be superior to 'black' races, through greater effort and struggle in the evolutionary race. Here, the supposed ascendancy of the 'white' races is shown explicitly, in Earnest Hooton's *Up from the ape*, published in 1932.

remarkably conservative, often remaining unchanged for several million years. In this light it is wrong to see evolution, or for that matter human history, as a constant progression, slow or otherwise.'

It is true that some species later in evolutionary time are more complex in certain ways than many earlier in time. But this can be explained simply as the ratchet effect, the fact that evolution builds on what existed before. For the most part, however, the world has not become a strikingly more complex place biologically as a whole: most organisms

remain simple, yet we are blinded by the exceptions, particularly the one with which we are most familiar.

Even this brief historical sketch clearly illustrates the anthropocentric spectacles through which paleoanthropologists have viewed the natural world in which we evolved. Such a perception is probably inescapable in some degree, as Glynn Isaac's earlier remark implied. In 1958, for instance, Julian Huxley, grandson of Thomas Henry, suggested that mankind's special intellectual and social qualities should be recognized formally by assigning *Homo sapiens* to a new grade, the Psychozoan. 'The new grade is of very large extent, at least in magnitude to all the rest of the animal Kingdom, though I prefer to regard it as covering an entirely new sector of the evolutionary process, the psychosocial, as against the entire non-human biological sector.'

The ultimate issue is 'the long-held view that humans are unique, a totally new type of organism', as Cambridge University's Robert Foley points out. This type of thinking leads to the notion that human origins therefore 'requires a special type of explanation, different from that used in understanding the rest of the biological world.'

Key questions:

● Did the intellectual framework provided by the Great Chain of Being lead naturally to the idea of the evolution of species?
● Why did Man's place in nature not change much in some ways between pre- and post-Darwinian eras?
● Why has the notion of progress become such an integral part of evolutionary thinking within Western philosophy, particularly in relation to human evolution?
● Does the evolution of qualitatively novel characteristics require qualitatively novel explanations?

Key references:

Donald K. Grayson, *The establishment of human antiquity*, Academic Press, 1983
Stephen Jay Gould, 'Vision with a vengeance', in *Natural history*, September 1980, pp 16–20; 'Bound by the great chain', *ibid*, November 1983, pp 20–24; 'Chimp on a chain', *ibid*, December 1983, pp 18–26.
Arthur O. Lovejoy, *The great chain of being*, Harvard University Press, 1936 (new reprint).

2 / Human evolution as narrative

Sir Grafton Elliot Smith: A leading anatomist and anthropologist in early twentieth-century Britain, Elliot Smith often wrote in florid prose about human evolution: see figure on next page. (Courtesy of University College, London.)

'One of the species specific characteristics of *Homo sapiens* is a love of stories,' noted the late Glynn Isaac, 'so that narrative reports of human evolution are demanded by society and even tend toward a common form.' Isaac was referring to the recent work of Boston University anthropologist Misia Landau, who has analyzed the narrative component of professional—not just popular—accounts of human origins.

'Scientists are generally aware of the influence of theory on observation', concludes Landau. 'Seldom do they recognize, however, that many scientific theories are essentially narratives.' This comment covers all sciences, but Landau identifies several elements in paleoanthropology that make it particularly susceptible to being cast in narrative form, both by those who tell the stories and by those who listen to them.

First, in seeking to explain human origins, paleoanthropology is apparently faced with a sequence of events through time that transformed an ape into a human. The description of such a sequence falls naturally into narrative form. Second, the subject of that transformation is ourselves. Being egotistical creatures, we tend to find stories about ourselves more interesting than stories about, for instance, bond energies in aliphatic hydrocarbons or alternative splicing routes in messenger RNA molecules.

Traditionally, paleoanthropologists have recognized four key events in human origins: terrestriality, bipedality, encephalization, and culture (or civilization). While these four events have usually featured in accounts of human origins, the order in which they have been thought to have occurred has differed.

For instance, Henry Fairfield Osborn, director of the American Museum of Natural History in the early decades of this century, considered the order was as given above, which, incidentally, coincides closely with Darwin's view. Sir Arthur Keith, a prominent figure in British anthropology in the 1920s, considered bipedalism to have been the first

event, with terrestriality following: in other words, Keith's ancestral ape took to walking on two legs while it was still a tree dweller; only subsequently did it descent to the ground. For Sir Grafton Elliot Smith, a contemporary of Keith's, encephalization led the way. His student, Frederick Wood Jones, agreed with Elliot Smith that encephalization and bipedalism developed while our ancestor still lived in trees, but thought that bipedalism preceded rather than followed brain expansion. William King Gregory, like his colleague Osborn, argued for terrestriality first, but differed in believing that the adoption of culture (tool use) preceded significant brain expansion. And so on.

So, here we see four common elements linked together in different ways, each scheme purporting to tell the story of human origins. And 'story' is the operative word. 'If you analyze the way in which Osborn, Keith and others explained the relation of these four events, you see clearly a narrative structure,' says Landau, 'but they are more than just stories. They conform to the structure of the hero folk tale.' In her analysis of paleoanthropological literature, Landau drew upon a system devised in 1925 by the Russian literary scholar Vladimir Propp. This system, published in Propp's

Morphology of the folk tale, included a series of 31 stages that encompassed the basic elements of the hero myth. Landau reduced the number of stages to nine, with the overall structure kept the same: hero enters; hero is challenged; hero triumphs.

In the case of human origins, the hero is the ape in the forest, who is 'destined' to become us: the climate changes, the forests shrink, and the hero is cast out on the savannah where he faces new and terrible dangers. He struggles to overcome them, by developing intelligence, tool use, and so on. And eventually emerges triumphant, recognizably you and I. 'When you read the literature you immediately notice not only the structure of the hero myth, but also the language', explains Landau.

For instance, Elliot Smith writes about '...the wonderful story of Man's journeyings towards his ultimate goal...' and '...Man's ceaseless struggle to achieve his destiny'. Roy Chapman Andrews, a colleague of Osborn's at the American Museum, writes of the pioneer spirit of our hero: 'Hurry has always been the tempo of human evolution. Hurry to get out of the primordial ape stage, to change body, brains, hands and feet faster than it had ever been done in the history of creation. Hurry on to the time when man could conquer the land and the sea and the air; when he could stand as Lord of all the Earth.'

Osborn himself wrote in similar tone: 'Why, then, has evolutionary fate treated ape and man so differently? The one has been left in the obscurity of its native jungle, while the other has been given a glorious exodus leading to the domination of earth, sea, and sky.' Indeed, Osborn's writings frequently explicitly embodied the notion of drama: 'The great drama of the prehistory of man...' and 'the prologue and opening acts of the human drama...', and so on.

Of course, it is possible to tell stories with similar gusto about the 'triumph of the reptiles in conquering the land', 'the triumph of birds in conquering the air', and so on. Such stirring tales are indeed readily to be found in accounts of evolutionary history—look no further than every child's hero, the dinosaur. But the fact that in paleoanthropology, the hero of the tale is *Homo sapiens*—ourselves—makes a significant difference. Although dinosaurs may be lauded as lords of the land in their time, only humans have been regarded as the inevitable product of evolution, indeed, the ultimate purpose of evolution, as we saw in the previous unit. Not everyone was as explicit about this as

Broom was, but most authorities betrayed the sentiment in the hero worship of their prose.

But these stories were not just accounts of the ultimate triumph of our hero. They carried a moral tale too: namely, triumph demands effort. 'The struggle for existence was severe and evoked all the inventive and resourceful faculties and encouraged [Dawn Man] to the fashioning and first use of wooden and then stone weapons for the chase', wrote Osborn. 'It compelled Dawn Man...to develop strength of limb to make long journeys on foot, strength of lungs for running, and quick vision and stealth for the chase.'

According to Elliot Smith, our ancestors '...were impelled to issue forth from their forests, and seek new sources of food and new surroundings on hill and plain, where they could obtain the sustenance they needed'. The penalty for indolence and lack of effort was plain for all to see, because the apes had fallen into this trap: 'While man was evolved amidst the strife with adverse conditions, the ancestors of the Gorilla and Chimpanzee gave up the struggle for mental supremacy because they were satisfied with their circumstances.'

Incidentally, in the literature of Elliot Smith's time—and until quite recently, albeit in more tempered form—the apes were usually viewed as

...xperieɩ.
. ɩɩe tremendous drama that .
.ɩɩis laboratory of mankind is based o.
.om a skull-cap and femur from Java, a sɩ.
ɩibia from Rhodesia, and an assortment of bones .om Western Europe!

But if we know nothing of the wonderful story of Man's journeyings toward his ultimate goal, beyond what we can infer from the flotsam and jetsam thrown upon the periphery of his ancient domain, it is essential, in attempting to interpret the meaning of these fragments, not to forget the great events that were happening in the more vitally important central area—say from India to Africa—and whenever a new specimen is thrown up, to appraise its significance from what we imagine to have been happening elsewhere, and from the evidence it affords of the wider history of Man's ceaseless struggle to achieve his destiny.

Nature has always been reluctant to give up to Man the secrets of his own early history, or, perhaps, uɩ ' ' ·ɩsiderate of his vanity in sparing him the fu'' '·se less attractive meɩɩɩʰ--- ᴏf '·ɩ retained ·'

Adventures in anthropology: Here, a short passage from Sir Grafton Elliot Smith's *Essays on the evolution of man*, published in 1924, illustrates the storytelling tone in which anthropological writing was often couched. Even modern prose is not always entirely free of this influence.

1 Initial situation   3 Change   5 Struggle/test   7 Transformation   9 Triumph !

2 Hero introduced   4 Departure   6 (Donor)   8 Tested again

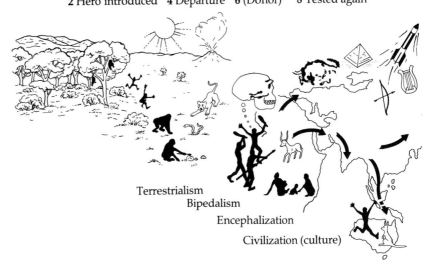

Terrestrialism
Bipedalism
Encephalization
Civilization (culture)

The hero-myth framework: Like folk tales ancient and modern, accounts of human origins have often followed the structure of hero myth. The hero (an ancient ape) sets off on a journey, during which he faces a series of challenges and opportunities that shape his final triumph (civilization). Recounting the evolution of any species is of course to tell a tale of a series of historical events: the effect, in the case of *Homo sapiens*, is to see the events as if, from the beginning, the journey was inevitable. (Courtesy of Misia Landau.)

evolutionary failures, left behind in the evolutionary race. The general perception until quite recently that the features of modern apes were evolutionarily primitive while those of hominids were advanced was a more scientific version of this same theme.

Although modern accounts of human origins usually avoid purple prose and implicit moralizing, there is one aspect of the narrative structure that lingers. This is the description of the events in the 'transformation of ape into human' as if each event were somehow a preparation for the next. 'Our ancestors became bipedal in order to make and use tools and weapons....tool-use enabled brain expansion and the evolution of language...thus endowed, sophisticated societal interactions were finally made possible...' Crudely put, to be sure, but this kind of reasoning was common in Osborn's day and is not unknown now.

Why does it happen? 'Telling a story does not consist simply in adding episodes to one another', explains Landau. 'It consists in creating relations between events.' Take, for instance, our ancestor's supposed 'coming to the ground', the first and crucial advance on the long road towards becoming human. It is easy to imagine how such an event can be perceived as a courageous first step on the long journey to civilization: defenseless ape faces the unknown predatory hazards of the savannah. 'There is nothing inherently transitional about the descent to the ground, however momentous the occasion', says Landau. 'It only acquires such value in relation to our overall conception of the course of human evolution.'

If evolution were steadily progressive, a program of constant improvement, the transformation of ape to human could be viewed as a series of novel adaptations, each one naturally preparing for and leading to the next. Such a scenario would be one of continual progress through time, and going in a particular direction. From our vantage point of the present, where we can view the end product, it is tempting to view the process in that way: because we can actually see that all those steps did in fact take place. But this ignores the fact that evolution tends to work in a rather halting, unpredictable fashion, shifting abruptly from one 'adaptive plateau' to another. And the world we have is simply a contingent fact, one of countless possibilities (see unit 4).

For instance, one cannot say that the first bipedal ape would inevitably become a stone-tool maker. In fact, if the current archeological record is any guide, those two events—bipedality and the advent of stone-tool making—were separated by at least 2.5 million years, and maybe much more. Another million years were to pass before further brain expansion occurred, again abruptly. The origin of anatomically modern humans after another million years or so was again an abrupt event. So, although it is often written than our ancestors were being propelled inexorably along an evolutionary trajectory, finishing up with *Homo sapiens*, that is simply describing what did happen: it ignores the many other possibilities that did not. As Landau remarks: 'There is a tendency in theories of hominid evolution to define origins in terms of endings.'

Darwin

Terrestriality Bipedalism Encephalization Civilization

Keith

Elliot Smith

Wood Jones

Osborn

Gregory

For paleoanthropology, language is an important scientific tool: it is used for the technical description of fossils and for the serious explication of evolutionary scenarios. It is true that all scientists should step back and scrutinize the language they use, because intertwined within it will be the elements of many unspoken assumptions. For paleoanthropology, where narrative becomes a particularly seductive vehicle for assumptions, it is especially important that from time to time one carefully examines what one says and the way one says it.

Key questions:

● What is implied by the fact that in Osborn's time, although paleoanthropologists employed the same set of events to describe the transformation of ape to human, those events were linked in many different combinations?
● Is paleoanthropology particularly susceptible to the invocation of the hero myth?
● Why do evolutionary scenarios in general lend themselves to narrative treatment?
● In what context were apes considered to be evolutionary failures?

Key references:

Niles Eldredge and Ian Tattersal, *The myths of human evolution*, Columbia University Press, 1982.
Glynn Isaac, 'Aspects of human evolution', in *Evolution from molecules to men*, Cambridge University Press, 1983, pp 509–543.
Misia Landau, 'Human evolution as narrative', *American Scientist*, vol 72, pp 262–268 (1984).
Misia Landau, 'Paradise lost: the theme of terrestriality in human evolution', in *The rhetoric of the human sciences*, edited by John S. Nelson, Allan Megill, and Donald N. McClosky, University of Wisconsin Press, 1987, pp 111–124.
Peter Medawar, *Pluto's Republic*, Oxford University Press, 1984.

Different views of the story: Even though anthropologists saw the human journey as involving the same fundamental events — terrestriality, bipedalism, encephalization, and civilization — different authorities sometimes placed these in slightly different orders. For instance, although Charles Darwin envisaged an ancient ape first coming to the ground, and then developing bipedalism, Sir Arthur Keith believed that the ape became bipedal before leaving the trees. (Courtesy of Misia Landau/*American Scientist*.)

3 / Historical views

Debate over human origins has advanced substantially in recent years, particularly in broadening the scientific basis of the discussions. Nevertheless, many of the issues addressed in current research have deep historical roots. A brief sketch of the subject's progress during the past 100 years or so will therefore put modern debates in historical context.

Two principal themes have been recurrent in this century of paleoanthropology, each of which has been seen to be more or less important at different times, depending on the ebb and flow of intellectual tides. First is the relationship between humans and apes: how close, how distant? And second is the 'humanness' of our direct ancestors, the early hominids.

During the past century, the issue of our relatedness to the apes has gone full cycle. From the time of Darwin, Huxley, and Haeckel until soon after the turn of the century, humans' closest relatives were regarded as being the African apes, the chimpanzee and gorilla, with the Asian great ape, the orangutan, being considered to be somewhat separate. Then, from the 1920s until the 1960s, humans were distanced from the great apes, which were said to be an evolutionarily closely-knit group. Since the 1960s, however, conventional wisdom has returned to its Darwinian cast.

This shift of opinions has, incidentally, been paralleled by a related shift in ideas on the location of the 'cradle of mankind'. Darwin plumped for Africa; Asia became popular in the early decades of the twentieth century; Africa is once again the focus.

While this human/African ape wheel has gone through one complete turn, the question of the humanness of the hominid lineage has been changing too, but in one direction only. Specifically, hominids—with the exception of *Homo sapiens* itself—have been gradually getting less human-like in the eyes of paleoanthropologists, particularly so in the last two decades.

Once Darwin's work finally established evolution as part of mainstream nineteenth century intellectual life, scientists had to account for human origins in naturalistic rather than supernatural terms. More importantly, as we saw from the previous unit, they also had to account for the very special qualities of humankind, those that appear to separate us from the world of nature. This was the challenge—and the response to it set the intellectual tone for a very long time to come.

In his *Descent of Man*, Darwin identified those characteristics that apparently make humans special—intelligence, manual dexterity and technology, uprightness—and argued that an ape endowed with just a little of each of these qualities would surely have an advantage over other apes. Once the earliest human forbear became established upon this evolutionary trajectory, the eventual

In the early decades of the twentieth century two opposing views of human origins were current:

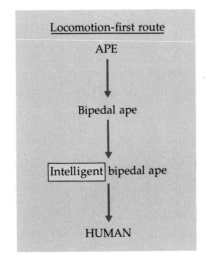

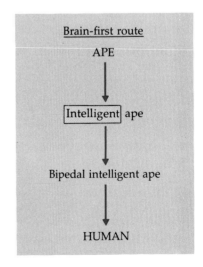

Conflicting views: One of the key differences of opinion regarding the history of human evolution was the role of the expanded brain: was it an early or a late development? The 'brain-first' notion, promoted by Elliot Smith, was important in paving the way for the acceptance of the Piltdown man fraud.

emergence of *Homo sapiens* becomes almost inevitable, through the continued power of natural selection. In other words, hominid origins become explicable in terms of human qualities. And *hominid* origins therefore equates with *human* origins. It was a seductive formula, and one that persisted until quite recently.

At the turn of the century several interrelated intellectual debates were going on, one of which was the order in which the major anatomical changes occurred in the human lineage. One notion was that the first step on the road to humanity was the adoption of upright locomotion. A second was that the brain led the way, giving an intelligent but still arboreal creature. It was into this intellectual climate that the perpetrator of the famous Piltdown hoax—a chimera of fragments from a modern human cranium and an orangutan's jaw, both doctored to make them look like ancient fossils—made his play in 1912.

The Piltdown 'fossils' appeared not only to confirm that the brain did indeed lead the way, but also that something close to the modern sapiens form was extremely ancient in human history. The

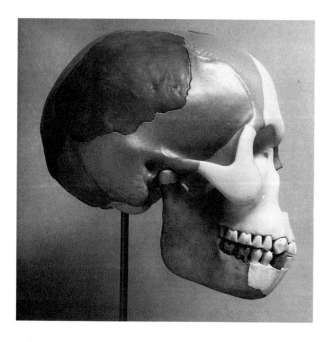

A fossil chimera: Cast of Piltdown reconstruction, based on lower jaw, canine tooth and skull fragments (shaded dark). The ready acceptance of the Piltdown forgery—a chimera of a modern human cranium and the jaw of an orangutan—derived from the British establishment's adherence to the brain-first route. (Courtesy of the American Museum of Natural History.)

apparent confirmation of this latter fact—extreme human antiquity—was important to the prominent British anthropologist Sir Arthur Keith and to Henry Fairfield Osborn, director of the American Museum of Natural History during the first three decades of the century, because their theories demanded it. One consequence of Piltdown was that Neanderthal—one of the few genuine fossils of the time—was disqualified from direct ancestry to *Homo sapiens*, because it apparently came later in time than Piltdown and yet was more primitive.

For Osborn, Piltdown was strong support for his Dawn Man theory, which stated that mankind originated on the high plateaus of Central Asia, not in the jungles of Africa. During the 1920s and 1930s, Osborn was locked in constant but gentlemanly debate with his colleague, William King Gregory, who carried the increasingly unpopular Darwin/Huxley/Haeckel torch for a close relationship between humans and African apes—the Ape Man theory.

Although Osborn was never very clear about what the earliest human progenitors might have looked like, his ally Frederick Wood Jones had firm ideas. Wood Jones, a British anatomist, interpreted key features of ape and monkey anatomy as specializations that were completely absent in human anatomy. He therefore proposed in 1919 his 'tarsioid hypothesis', which looked for human antecedents very low down in the primate tree.

In today's terms, Wood Jones's proposal would put human origins in the region of 50 to 60 million years ago, close to the origin of the primate radiation, while Keith's notion of some kind of early ape would put it in the region of 30 million years ago.

During the 1930s and 1940s, the anti-ape arguments of Osborn and Wood Jones were lost, but Gregory's position did not immediately prevail. Gregory had argued for a close link between humans and the African apes on the basis of shared anatomical features. Others, including Adolph Schultz and D.J. Morton, said, however, that although humans probably derived from apelike stock, the similarities between humans and modern African apes were the result of parallel evolution. This position remained dominant through the 1960s, firmly supported by Sir Wilfrid Le Gros Clark, Britain's most prominent primate anatomist of the time. Humans, it was argued, came from the base of the ape stock, not way up with the specializations developed by the African apes.

A discussion of the Piltdown skull: Back row, left to right: F.G. Barlow, Grafton Elliot Smith, Charles Dawson, and Arthur Smith Woodward. Front row, left to right: A.S. Underwood, Arthur Keith (examining the skull), W.P. Pycraft, and Ray Lankester. The Piltdown man fossil, discovered in 1912 and exposed as a fraud in 1953, fitted so closely with British anthropologists' view of human origins that it was accepted uncritically as being genuine. (Courtesy of the American Museum of Natural History.)

During the 1950s and 1960s, fossil evidence of early apes accumulated at a significant rate, and it seemed to show that these creatures were not simply early versions of modern apes, as had been tacitly assumed. This meant that those authorities who accepted an evolutionary link between humans and apes, but did not accept a close human/African ape link, did not now have to go way back in the history of the group to 'avoid' the specialization of the modern species. At the same time, those who insisted that the similarities between African apes and humans were the result of common heritage, not parallel evolution, were forced to argue for a very recent origin of the human line. Prominent among proponents of this latter argument was Sherwood Washburn, of the University of California, Berkeley.

One of the fossil discoveries of the 1960s—in fact, a rediscovery—that appeared to confirm the notion of parallel evolution to explain human/African ape similarities was made by Elwyn Simons, then of Yale University. *Ramapithecus* was the fossil specimen, an apelike creature that lived in Eurasia about 15 million years ago and appeared to share many anatomical features (in the teeth and jaws) with hominids. Simons, later supported closely by David Pilbeam, proposed *Ramapithecus* as the beginning of the hominid line, thus excluding a human/African ape connection.

Arguments about the relatedness between humans and African apes took place against a rethinking about the relatedness among the apes themselves. In 1927, G.E. Pilgrim had suggested that the great apes be treated as a natural group, with humans evolutionarily more distant. The idea eventually became popular, and was the accepted wisdom until molecular biological evidence undermined it in 1963, the work of Morris Goodman at Wayne State University. Goodman's molecular biology data on blood proteins indicated that humans and the

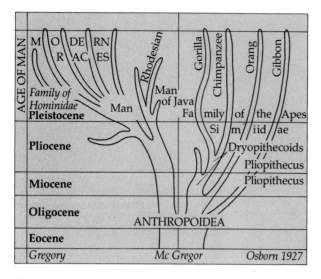

(a)

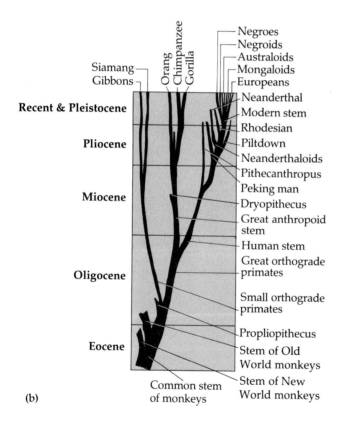

(b)

Two phylogenetic trees: Henry Fairfield Osborn's 1927 view of human evolution (a) shows a very early division between humans and apes—in todays geological scale, this division would be about 30 million years ago. (b) shows Sir Arthur Keith's slightly earlier rendition. Again, the human/ape division is seen as very early; and again, long lines link modern species with supposed ancestral stock, a habit that was to persist until quite recently. Note also the very long history of modern human races.

African apes formed a natural group, with the orangutan more distant.

Thus, the Darwin/Huxley/Haeckel position was reinstated, with first Gregory and then Washburn its champions. Subsequent molecular biological— and fossil—evidence seems to confirm Washburn's original suggestion that the origin of the human line is indeed recent, lying between 5 and 10 million years ago. *Ramapithecus* was no longer regarded as the first hominid, but simply one of many early apes.

Meanwhile, discoveries of fossil hominids, and the stone tools they apparently made, had been gathering pace during the 1940s through 1970s, first in South Africa and then in East Africa. Culture—specifically, stone-tool making and use

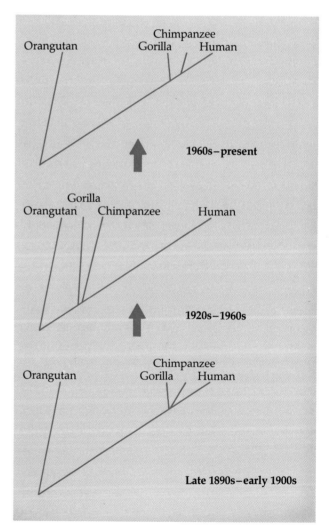

Shifting patterns: Between the beginning of the century and now, ideas over the relationships among apes and humans have gone full circle.

in hunting and butchering animals—became a dominant theme, so much so that to be hominid implied a hunter–gatherer lifeway. The importance of culture as *the* hominid characteristic became expressed at its most extreme in the single species hypothesis, promulgated during the 1960s principally by C. Loring Brace, of the University of Michigan.

Only one species of hominid existed at any one time, stated the hypothesis, human history progressing by the steady improvement up a single evolutionary ladder. The rationale rested upon a supposed rule of ecology: the principle of competitive exclusion. In this case, culture was viewed as such a novel and powerful behavioral adaptation, that two cultural species simply could not coexist. And since all hominids are cultural by definition, only one hominid species existed at any one time.

The single species hypothesis collapsed finally in the mid-1970s, with fossil discoveries from Kenya that indisputedly showed the coexistence of two very different species of hominid: one was *Homo erectus*, a large-brained species that apparently was ancestral to *Homo sapiens*; the other was *Australopithecus boisei*, a small-brained species that eventually became extinct. Subsequent discoveries and analyses implied that there might have been several species of hominid coexisting in Africa some 2 million or so years ago, suggesting that several different ecological niches were being successfully exploited. And this implied that to be hominid did not necessarily mean being cultural. So, no longer could hominid origins be equated with human origins.

During the past decade, not only has there been an appreciation of a spectrum of hominid adaptations—which includes the notion simply of a bipedal ape—but the lineage that eventually led to *Homo sapiens* has also come to be perceived as much less human. Gone is the notion of a scaled-down version of a modern hunter–gatherer way of life. In its place has appeared a rather unusual African ape adopting some novel, un-apelike modes of subsistence.

Hominid origins are therefore now completely divorced from any notion of human origins. Questions about the beginning of the hominid lineage are now firmly within the territory of behavioral ecology, and do not draw upon those qualities that we might perceive as separating us from the rest of

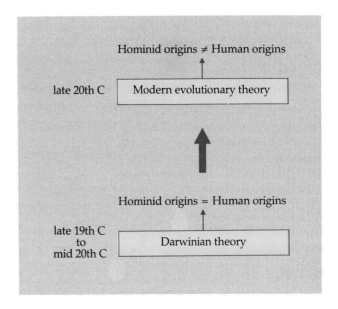

Hominids as humans: Until quite recently anthropologists frequently thought about humanlike characteristics while considering hominid origins, a habit that can be traced back to Darwin. Now, the humanity of hominids is seen as a rather recent evolutionary development.

animate nature. Questions of human origins have now to be posed within the context of primate biology.

Key questions:

● Why were post-evolutionary theory explanations of human origins 'self explanatory'?
● What is the effect of sparse fossil evidence on theories of human evolution?
● Was the notion of parallel evolution of similar anatomical features among humans and African apes a reasonable explanation?
● Why was 'culture' so dominant a theme in explanations of human origins?

Key references:

Peter J. Bowler, *Theories of human evolution*, Johns Hopkins Press, 1986.
Matt Cartmill, David Pilbeam, and Glynn Isaac, 'One hundred years of paleoanthropology', *American Scientist*, vol 74, pp 410–420 (1986).
John G. Fleagle and William L. Jungers, 'Fifty years of higher primate phylogeny', in *A history of American physical anthropology*, edited by Frank Spencer, Academic Press, 1982.
Roger Lewin, *Bones of contention*, Simon and Schuster, 1987.

4 / Modern evolutionary theory

'I had two distinct objects in view; firstly to show that species had not been separately created, and secondly, that natural selection had been the chief agent of change.' Thus did Darwin characterize his objectives in writing the *Origin of species*, which was published in November 1859. He succeeded immediately with the first, but the second took longer. Natural selection did not become regarded as an important engine of evolutionary change until the 1930s, and even now there is lively debate over the precise nature of evolutionary mechanisms.

The shift between pre- and post-Darwinian eras brought with it two important changes in the way biologists viewed the world. First, the idea of design, which in natural theology was said to be evidence of God's hand, was replaced by the concept of *adaptation*, organisms becoming fitted to their environment through naturalistic processes. Second, the notion of some kind of meaning in the overall pattern in nature—the Great Chain of Being, for instance—was replaced by *historical perspective*, that the pattern we see is the product of descent.

The phenomenon of adaptation is what any successful theory of evolution has to explain, and

Darwin's key innovation was the theory of natural selection (invented simultaneously with another British natural historian, Alfred Russell Wallace). The concept of natural selection is simple, and embodies two key steps: first, within a population there is heritable, phenotypic variation; second, as a result of that variation, there is differential reproductive success.

As all species produce more offspring than will eventually survive, any slight advantage enjoyed by a particular parent's offspring as a result of a heritable trait will lead to greater representation in the next generation. And generation by generation the trait will spread through the population, leading to evolutionary change.

Variation and selection: a simple two-stage process. But, key to Darwin's natural selection is that the source of the variation should be undirected. Darwin viewed natural selection as a creative process, building adaptation step by step. This works only if the source of variation is not directed by the environment (or by the organism, which is at the heart of Lamarckism), so that selection is then favoring the advantageous traits among that 'random' variation. By contrast, if new environments could elicit favorable variation within a population, then the process of selection that follows would simply be to eliminate the unfit.

Darwin's arguments on natural selection failed to convince many of his contemporaries, partly because, until the turn of the century, there was no adequate theory of genetic inheritance: heritable variation could therefore not be fully understood. In addition,

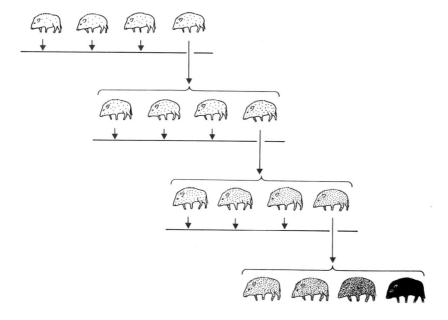

Natural selection: Selective advantages of large body size and dark coat colour confer differential survival and ultimate reproductive success on those individuals with those characters. As a result, the population mean for these characters will shift, generation by generation, towards larger size and darker coats.

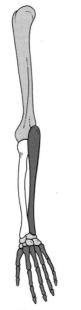

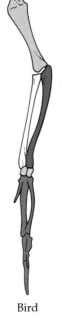
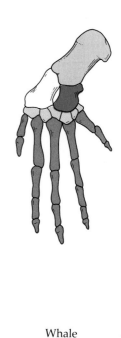

The principle of homology: The biological derivation relationship (shown by colours) of the various bones in the forelimbs of four vertebrates is known as homology and was one of Darwin's arguments in favour of evolution. By contrast, the wing of a bird and the wing of a butterfly, although they do the same job, are not derived from the same structures: they are examples of analogy.

Human   Dog   Bird   Whale

however, Darwin's message was unpalatable: evolution, he seemed to be saying, was purposeless, without some higher guiding principle. Moreover, he appeared to be saying that life was not inherently progressive, moving step by step up a ladder of improvement. As Harvard biologist Stephen Jay Gould notes: 'No misunderstanding of evolution is more widespread than the assumption that it inevitably leads to a progressive improvement of life.'

When Gregor Mendel's work on genetic inheritance was rediscovered at the turn of the century, the science of genetics was born. And with it came an emphasis on the power of mutation to direct evolutionary change, with selection assigned simply to eliminating the unfit.

By the 1930s, however, the intellectual tide was turning, and the essence of natural selection began to be appreciated. But the tide took a decade to turn fully, marked finally by the advent of the 'modern synthesis', an amalgam of population genetics, natural history, systematics, paleontology, and morphology, centered on Darwinian natural selection. For two decades the modern synthesis, or NeoDarwinism as it is also called, was intellectually vibrant, with the perceived link between natural selection and adaptation becoming ever tighter. The gradual nature of evolution by natural selection was increasingly emphasized, as was its power in shaping adaptations.

Indeed, the literature of the time came to imply that all of evolution—including trends seen in the fossil record, such as the change in size and form of the horse—was achieved gradually, and that

more or less any every aspect of an organism was the outcome of adaptation. Doubts began to appear in the 1970s, both about the gradual nature of change and about the omnipotence of adaptation.

First, it was discovered that most populations contained more genetic variation within them than was predicted under strict selectionism. Other mechanisms appeared to be operating, including what has come to be known as neutral evolution. In addition, some critics argued that steady, gradual change was the exception, not the rule, through long periods of time. Change was concentrated in geological brief periods of time, they said, with species remaining unchanging—in stasis—for most of their existence. (The periods of change might measure 50,000 years, for instance, with stasis extending to 2 million years.)

Evolutionary trends were therefore seen to be the result of the differential production of species with certain characteristics, not gradual change within lineages as implied by NeoDarwinism. This view of evolutionary history, which came to be known as 'punctuated equilibrium', was first proposed in 1972 by Niles Eldredge and Stephen Jay Gould. Lively debate has ensued ever since on its relative merits.

While strict NeoDarwinians highlighted the power of natural selection in building adaptations through gradual change, punctuationists emphasized the constraints on change. Specifically, while accepting a fundamental role for natural selection, punctuationists pointed out that the routes of selection are constrained by inherited morphology.

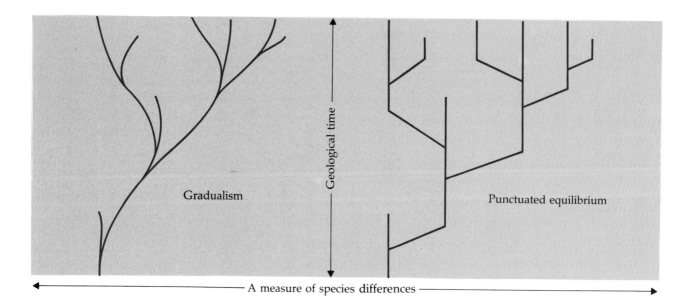

Geological time

A measure of species differences

Two modes of evolution: Gradualism and punctuated equilibrium. Gradualism views evolution as proceeding by the steady accumulation of small changes over long periods of time. Punctuated equilibrium, by contrast, sees morphological change as being concentrated in 'brief' bursts of change, usually associated with the origin of a new species. Evolutionary history is the outcome of a combination of these two modes of change; however, there is considerable debate as to which mode is the more important.

As the French biologist Francois Jacob put it, evolution works like a tinkerer—it works with what it's got, not like an engineer, who would start from scratch. Secondly, changing morphology in one part of the body usually causes correlated changes in another part, which have nothing to do with adaptation.

Fundamentally, these constraints relate to the limited number of pathways by which embryonic development can operate, both in absolute terms (there are no animals with wheels, for instance), and at any point in the history of a particular lineage. NeoDarwinians were not unaware of such constraints, it was just that they had become under-emphasized in the intellectual success of strict selectionism.

In this same context, one thing that has become very clear is that there is no simple relationship between genetic change—mutation—and the degree of phenotypic change it might produce. For instance, a single base mutation in the gene of a serum albumin might marginally modify the physical chemistry of the blood, but with no significant impact on adaptation. However, a similar mutation in a gene that affects the timing of the program of embryological development might have dramatic consequences for the mature organism. The slowing of embryological development, with prolongation of the growth period, is apparently important in

the evolution of humans from apes: the phenomenon is known as neotony.

This 'uncoupling' of magnitude of mutation from magnitude of phenotypic change has clear consequences for the ability to infer genetic distance from degree of morphological difference.

When he wrote the *Origin of species*, Darwin argued that the reason that 'intermediate forms' were rare in the fossil record was the extreme incompleteness of the record: brief snapshots at long-separated intervals through time. The theory of punctuated equilibrium, while not denying that the record is incomplete, gives another interpretation: namely, that evolutionary shifts are concentrated in geologically brief periods of time and in small, peripheral populations. There is therefore little opportunity for such populations to be incorporated in the record. In other words, the pattern seen in the record—species persisting in one form, then abruptly changing to another—is a reflection of reality, not an artifact of the record itself.

The generation of new species is of course central to evolutionary biology theory, and several issues surround it, some of which are intertwined with the gradualist/punctuationist debate. One issue is the modes by which new species may arise; a second is the circumstances under which this is most likely to occur.

First, the modes of origin, of which there are

two: phyletic evolution and speciation. In phyletic evolution an entire species becomes transformed gradually over time, becoming so different that it warrants being called a new species, the ancestral species effectively having become extinct. In speciation a subpopulation splits off from the ancestral population and changes sufficiently to become a new species, the ancestral population continuing to exist. With speciation, therefore, there is branching, yielding two species where previously there was one. Very roughly, gradualists concentrate on phyletic transformation while punctuationists emphasize speciation, although such a division by no means exclusive.

Speciation has obviously been important in the history of life, because it has more than balanced the steady 'background' rate of extinction and the occasional 'mass' extinctions: species diversity has on average increased through time. Today there are estimated to be some 30 million species, most of which are insects, many are plants, some 8600 are birds, and just 4000 mammals. In historical terms, this comprises about 1 per cent of all the species that have ever existed. The typical 'lifespan' for an invertebrate species is between 5 and 10 million years, and for a vertebrate more like 1 to 2 million years.

The circumstances that influence the likelihood that a new species might arise are many, and include characteristics of the ancestral species themselves as well as environmental factors.

The effect of population size on the origin of species has long been debated among evolutionary biologists. Are small populations more conducive to evolutionary change, because they escape the genetic inertia built into large, successful populations? Or are large populations more likely to generate evolutionary change, because the pool of genetic variation is so great? Punctuationists tend to favor the former, gradualists the latter.

No one denies that small populations can give rise to new species, such as when a small number of individuals colonize an island. In this case—known as a founder effect—the colonists possess only a proportion of the genetic variation present in the original population, but because they are isolated can undergo surprisingly dramatic change, effectively producing a new species. Genetic elements that were present in low frequency in the original population might, if carried in the colonists, become important in the founder population. A similar effect in terms of reduced initial variation

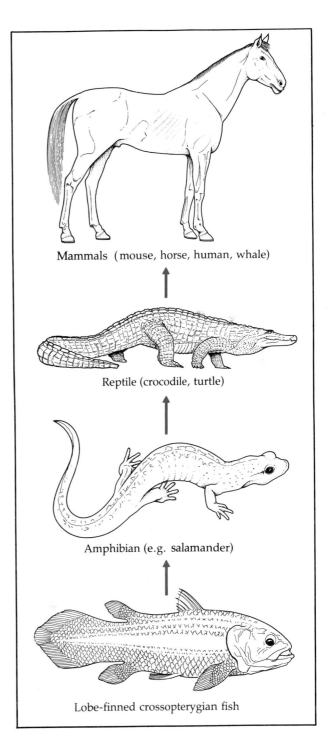

Mammals (mouse, horse, human, whale)

Reptile (crocodile, turtle)

Amphibian (e.g. salamander)

Lobe-finned crossopterygian fish

The principle of historical constraint: Evolution, in many ways, is a conservative process. The preservation of a four-limbed body over vast tracts of time and through very different environmental circumstances illustrates the power of historical constraint. For example, the horse has four legs not just because it is a very efficient way of moving about on dry land but because its fish ancestors also had four appendages.

followed by new genetic arrangements is produced when a large population 'crashes' to a small number of individuals: the phenomenon is known as a bottleneck.

Small populations of an otherwise widespread species can be produced by less dramatic means, of course, such as a group of individuals becoming isolated by a geographical barrier such as a mountain range or a river. Again, the same potential for genetic change and speciation arises. The question is, how frequent does this type of situation arise as against selection producing evolutionary change within a large population, acting on the large pool of genetic variation present in the population? At present there is no agreement among evolutionary biologists on this point.

A second factor affecting potential speciation is the nature of a species' adaptation. The fossil record shows that species that have highly specialized environmental and subsistence requirements are more likely to speciate than those with much broader adaptations. The reason is that any change in prevailing environment is likely to push specialists to the limits of their tolerances, promoting both speciation and extinction. Clearly, generalists can accommodate much broader shifts in conditions, making speciation and extinction rarer for them.

Key questions:

- Why must natural selection be seen as a creative force in relation to adaptation?
- What does the existence of non-adaptive traits tell one about the nature of evolutionary change?
- In addition to a narrow adaptive niche, what other traits might make a species more susceptible to speciation and extinction?
- What explanations might there be for hominids being an unusually species-poor group?

Key references:

Stephen Jay Gould, 'Darwinism and the expansion of evolutionary theory', *Science*, vol 216, pp 380–387 (1982).

Francois Jacob, 'Evolution and tinkering', *Science*, vol 196, pp 1161–1166 (1977).

Motoo Kimura, 'The neutral theory of molecular evolution', *New Scientist*, 11 July 1985, pp 41–44.

G. Ledyard Stebbins and Francisco J. Ayala, 'The evolution of Darwinism', *Scientific America*, July 1985.

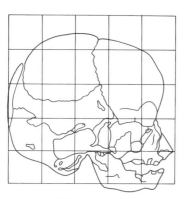

Chimp fetus

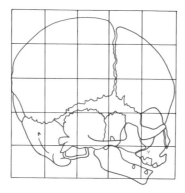

Human fetus

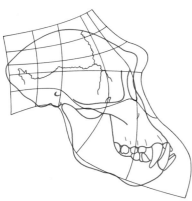

Chimp adult

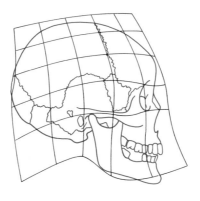

Human adult

Neotony in human evolution:
Although the shape of the cranium in human and chimpanzee fetuses is very similar, a slowing down in development through human evolution has produced adult crania of very different forms, principally in the shape of the face and the size of the brain case. The changes in grid shapes indicate the orientation of growth.

5 / Geology and climate drive evolution

'It is probably reasonable to conclude that, had it not been for temperature-based environmental changes in the habitats of early hominids, we would still be secure in some warm hospitable forest, as in the Miocene of old, and we would still be in the trees.' This speculation, by C.K. Brain of the Transvaal Museum in Pretoria, South Africa, encapsulates the notion that environment—particularly environmental change—has been an important engine driving human evolution.

Since Brain made his speculation in 1981, a veritable research industry has geared itself up in an attempt to test the idea. Geologists, climatologists, and paleontologists have combined their different datasets—and are beginning to see a distinct pattern emerge from the points. Briefly, that pattern indicates that episodes of significant global cooling are accompanied by pulses of extinctions and speciations among the world's biota. And it appears that hominids are no exception to this evolutionary pattern.

NeoDarwinism has always held the environment to be central to evolutionary change: different environments demand different adaptations. A species' environment is basically of two kinds: the physical world and the resources within it; and other species with which it interacts, specifically with which it competes in 'the struggle for existence' (Darwin's phrase).

Some evolutionary biologists argue that even in the absence of change in the physical environment, species would be constantly evolving, driven by competition with other species. Indeed, the Red Queen hypothesis, as this point of view is known, argues that interspecific competition is the principal driving force of evolution. In general, however, most investigators regard the physical environment as an important—if not the prime—engine initiating evolutionary change.

Changes in physical environment has influenced the history of life on at least three levels of scale, some of which are interlinked: extraterrestrial, global geography, and local climate.

Geologists have long known that Earth history is punctuated by mass extinction, the biggest being the Permian extinction 220 million years ago, when about 95 per cent of all species apparently perished. Five such mass dyings are known, together with a series of somewhat smaller ones, which, according to some authorities, occur regularly, at intervals of about 27 million years, the last one being 13 million years ago. The hypothesis of periodic extinction remains highly controversial.

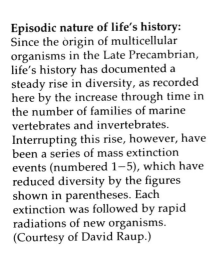

Episodic nature of life's history:
Since the origin of multicellular organisms in the Late Precambrian, life's history has documented a steady rise in diversity, as recorded here by the increase through time in the number of families of marine vertebrates and invertebrates. Interrupting this rise, however, have been a series of mass extinction events (numbered 1–5), which have reduced diversity by the figures shown in parentheses. Each extinction was followed by rapid radiations of new organisms. (Courtesy of David Raup.)

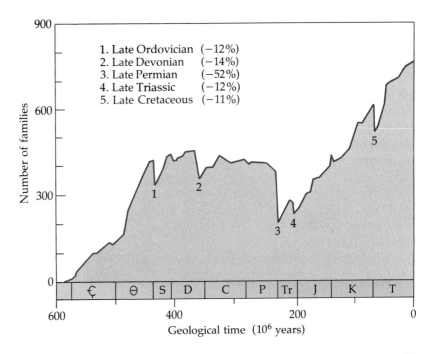

1. Late Ordovician (−12%)
2. Late Devonian (−14%)
3. Late Permian (−52%)
4. Late Triassic (−12%)
5. Late Cretaceous (−11%)

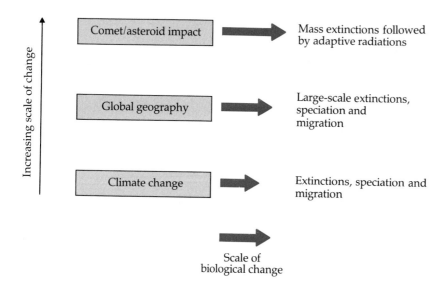

Increasing scale of change

Comet/asteroid impact → Mass extinctions followed by adaptive radiations

Global geography → Large-scale extinctions, speciation and migration

Climate change → Extinctions, speciation and migration

Scale of biological change

Scales of change: Habitats can be altered by external changes, either directly through climate, less directly through geography, and catastrophically through asteroid or comet impact. The most common response to habitat change is migration, with plant and animal species following their preferred climate: in cooling trends, migration in the northern hemisphere will be southwards, in warming trends, northwards. When habitat fragmentation prevents migration, species may become extinct or undergo speciation.

In recent years the scientific community has taken very seriously the suggestion that the Cretaceous/Tertiary extinction 65 million years ago, which drew an end to the age of dinosaurs, may have been caused by the impact of a comet or asteroid with the earth. And some of those researchers who accept the notion of periodic extinction have suggested that the cause is periodic collision of the earth with comets or asteroids.

Whatever the cause of mass extinctions, their effect is devastating. Whole groups may be wiped out or greatly reduced in diversity. Following the extinction event, the biota rebounds, many groups going through vigorous adaptive radiations—the placental mammals, following the Cretaceous/Tertiary extinction, are a good example. Clearly, the pattern of rebound and subsequent history of life depends on which lineages survive. It seems that the impact of mass extinctions is so great that, except for certain characteristics such as degree of geographic distribution, lineage survival is largely a matter of good luck. It is salutary to remember that the Cretaceous/Tertiary extinction spelled the end for many groups of placental mammals. The primates just happened to be among those that survived.

On the next level of scale is tectonics, the constant movement of the dozen or so major plates that constitute the earth's crust and upon which the continents ride. Biotas that were once united have been divided, and previously independent biota have been brought together.

For instance, Old World and New World monkeys derive from a common stock, but followed independent paths of evolution as South America and Africa drifted apart some 60 million years ago. Australia's menage of marsupial mammals evolved in isolation from placental mammals, the island continent having lost contact with Old World landmasses more than 50 million years ago, before placental mammals invaded.

By contrast, when the Americas joined some 3 million years ago there was a massive exchange of biotas that had evolved separately for tens of millions of years. Indian and Asian species migrated into each other's lands when the continents united some 45 million years ago. India's continued northward movement eventually caused the uplift of the massive Himalayan range, producing further geographic and climatic modification on a grand scale. Africa and Eurasia exchanged species when the landmasses made contact about 18 million years ago, apes being among those species making the journey from south to north.

Whenever landmasses become isolated as a result of plate tectonics, the environment—and therefore the evolutionary fate—of the indigenous species was influenced simply by the fact of isolation. More dramatic, however, was the effect of uniting previously separated landmasses, because it brought a combination of new opportunities and the hazard of new competition to the biotas. Some groups diversified in these circumstances, as did the apes as they spread out of Africa, while others succumbed to extinction, the fate of many South American mammals during the Great American Interchange.

In addition to influencing evolution by shuffling landmasses, plate tectonics can also modify the environment within continents, a prime example

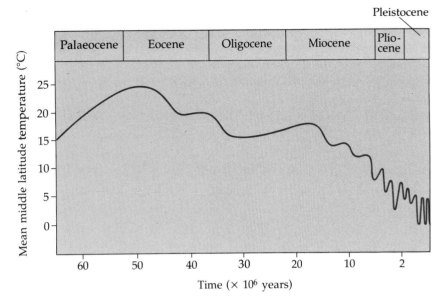

Global temperature changes: For reasons not completely understood, average global temperature has decreased during the past 50 million years. Although polar ice probably began forming as much as 30 million years ago, it has been an important modulator of global climate only in the last 10 million and particularly the last 5 million years. Temperature fluctuations have an important impact on habitat migration and fragmentation.

of which occurred in Africa. Crudely speaking, 20 million years ago the continent was carpeted west to east with tropical forest. Today, however, East Africa is a mosaic of savannah and open woodland, separated from the still continuous forest to the west by the Great Rift Valley.

A minor tectonic plate margin runs south to north under East Africa, the first consequence of which was 'doming' that began 15 million years ago, producing tremendous lava-driven uplifts reaching 1000 meters high and centered near Nairobi in Kenya and Addis Ababa in Ethiopia. Then, weakened by the separating plates, the continental rock collapsed in a long, vertical fault, snaking several thousand kilometers from Tanzania in the south to Ethiopia in the north. The effect of all this was to throw the eastern part of the continent into rain shadow, thus dramatically altering the vegetation. These tectonic processes were accompanied by episodes of global cooling, which accentuated the replacement of forest by more open environments. The combination must have been key to early hominid evolution, which appears to have taken place there.

Climate represents a third level of scale influencing species environment. Clearly, the nutrient resources exploited by a species will be influenced by the prevailing climate, and any change may affect a species' ability to survive in a particular locality. But, in terms of major evolutionary change, more important than resources is the integrity of the species' habitat as a whole. Specifically, any fragmentation of a species' habitat range as a result of significant climate shift may lead to speciation in some cases and extinction in others (see unit 4).

This, briefly stated, is the basis of the 'turnover-pulse' hypothesis advanced by Elisabeth Vrba, of Yale University.

The most common response of a species to changing climate is to migrate, following the conditions to which it is adapted: in the northern hemisphere, southward during times of cooling, and northward when conditions warm up. However, migration is not always possible, being prevented by physical barriers such as mountain ranges and rivers, or biological barriers, such as the absence of food resources and water. In such cases, populations may become fragmented and perhaps subject to different prevailing conditions. Too great a change, and extinction is likely. Moderate change, and speciation is possible.

Because episodes of significant cooling are likely to make northern latitudes less habitable, speciations will be concentrated in equatorial zones during such times. African lineages are therefore likely to enjoy a higher rate of speciation than those in Eurasia, because of the continent's equatorial location. 'Modern global patterns of species diversity in Bovidae are in accord with this prediction', notes Vrba. 'Subsaharan African alone has roughly twice as many endemic species as all of Eurasia.' Other considerations aside, Africa was therefore statistically more likely to have been the 'cradle of mankind' than was any other continent, simply by virtue of its position on the globe.

The turnover-pulse hypothesis states, therefore, that when lineages experience extinctions and speciations, they will do so synchronously and in coincidence with major climatic change, particularly cooling episodes. The hypothesis is still being tested, but so far the results appear to support it. The task is to identify major climatic shifts; to look for pulses of extinctions and speciations in the fossil record; and if such pulses exist, to see if they coincide with the climatic episodes. Because of the nature of the fossil record the best data come from marine sediments, while continental sequences lamentably are still fragmentary.

First, the climatic record, which centers on the formation of the polar ice caps as an indication of cooling episodes. During the first half of the Cenozoic period (65 million years ago to the present), which encompasses the history of the primate order, the globe appears to have been consistently free of significant polar ice. The first appearance of polar ice was the formation of the East Antarctic ice sheet a little after 35 million years ago, the Arctic remaining ice-free. The next climatic step

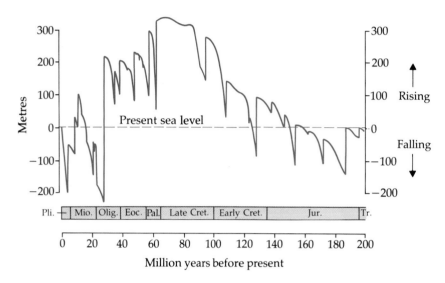

Sea level fluctuation: Sea level has fluctuated substantially over the past 200 million years, with one large cycle of rise and fall punctuated by short, rapid cycles. At least some of the precipitous drops are associated with the onset of glaciation, particularly in recent times (30 million years onwards). (Courtesy of P.R. Vail *et al.*/AAPG.)

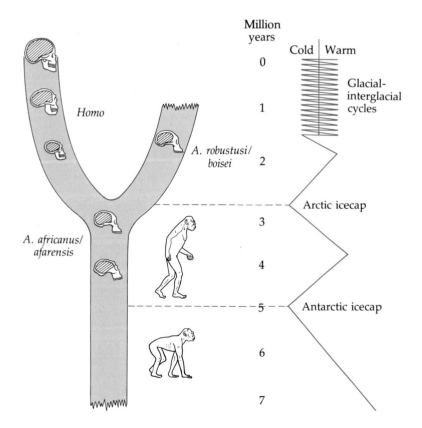

Climate and human evolution: Key events in human history appear to coincide with major climatic coolings, at 5 million, 2.5 million, and 0.9 million years ago. C.K. Brain's 1981 depiction of the causal connection between climate and human evolution was important in stimulating research on the link. (Courtesy of C.K. Brain.)

was the formation of the West Antarctic ice sheet, beginning soon after 15 million years ago.

Antarctic ice appears to have been a permanent global feature after this point, although it fluctuated in extent, with major advances at around 5 million and 2.4 million years, the latter being brief but massive. Significantly, 2.4 million years ago was also the date of the first appearance of Arctic ice. Another climatic pulse occurred 0.9 million years ago, which set in train the main Pleistocene glaciations.

Probably the best dataset of continental vertebrates is that of the African bovids (various kinds of antelope), much of which has been collected by Vrba. Although the data are not suitable to test the 15 million year climatic pulse, they clearly show spikes of extinctions and speciations at 5 million, 2.4 million, and something less than 1 million. The 2.4 million year spike is especially pronounced, and there are mammalian fossil data from Europe and Asia that also reflect what apparently was an extreme climatic episode.

What of the hominoids, including human ancestors? The 15 million year climatic episode does appear to coincide with a diversification of hominoids in Africa and Eurasia. But this expansion of species diversity also coincided with an expansion of geographic range following the joining of Africa with Eurasia. This one is therefore difficult to call. Towards the Late Miocene—8 to 5 million years ago—hominoids became extinct in Eurasia, coinciding both with local indications of changing environments and with the onset of the West Antarctic ice sheet advance. This was also the period during which, according to molecular biological evidence, the chimpanzee and hominid lineages differentiated.

The 2.4 million year event is close to the point of origin of pygmy chimpanzees, but is right on the mark for some people's estimate for the beginning of the genus *Homo*, and possibly the origin of two lineages of robust australopithecine. The robust australopithecines became extinct about a million years ago, which is close to the 0.9 million year climatic event, as too is the first expansion of hominids out of Africa.

The turnover-pulse hypothesis clearly cannot be tested by data from hominid history, because for that one needs speciose groups, such as the antelopes. But, inasmuch as certain climatic events appear to be real and appear to be tracked by speciations and extinctions in some mammalian groups, some light can be thrown on the initiation of speciation within the hominid group. Of the data available so far, some degree of confidence can be placed in the 2.4 million year event, both in climatic and evolutionary terms. In decreasing degrees of confidence come the events at 5, 0.9 and 15 million years.

Key questions:

- In what particular circumstances might competition between species drive constant evolutionary change?
- What does the pattern of mass extinctions imply about the evolutionary history of life on Earth?
- What types of fauna would be expected to be evolving in concert with the appearance of the hominids in East Africa towards the end of the Miocene?
- What types of data would falsify the turnover-pulse hypothesis?

Key references:

C.K. Brain, 'The evolution of man in Africa,' *The Geological Society of South Africa, Annexure to volume LXXXIV* (1981).

Richard A. Kerr, 'Ocean drilling details steps to an icy world,' *Science*, vol 236, pp 912–913 (1987).

Leo F. Laporte and Adrienne L. Zihlman, 'Plates, climate and hominoid evolution,' *South African Journal of Science*, vol 79, pp 96–110 (1983).

Elisabeth S. Vrba, 'Environment and evolution: alternative causes of the temporal distribution of evolutionary events,' *South African Journal of Science*, vol 81, pp 229–236 (1985).

6 / Systematics: or who is related to whom?

Ever since the time of Aristotle, scientists have classified organisms into groups, and the framework upon which modern biologists depend goes back to the mid-eighteenth century, when Carolus Linnaeus published his *Systema Naturae*. Classification—or the science of systematics—is important because it allows biologists to communicate with the minimum of ambiguity. And it has always been the ambition of biologists to classify organisms according to some kind of 'natural' system. The question is: what should that system be?

Before the 1859 publication of Darwin's *Origin of species*, classification clearly did not explicitly reflect evolution. After 1859, however, biologists could approach classification with evolution explicitly in mind, if it were considered appropriate. In fact, Darwin argued that because species were related by common descent, genealogy was the only logical basis for classification. Recent years have witnessed surprisingly heated debate over precisely how evolution should properly be incorporated into classification: should it emphasize the

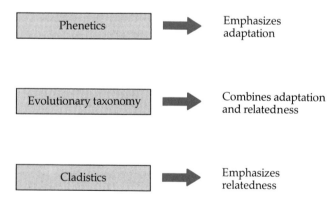

Approaches to classification: Different methods of describing relationships among organisms effectively emphasize different aspects of the world. For instance, by concentrating only characteristics that reflect genetic relatedness, cladistics produces an evolutionary tree. Phenetics, by contrast, measures all aspects of similarity among organisms, and therefore emphasizes similarities in adaption. Evolutionary taxonomy steers a middle path between the two.

results of evolution, in terms of adaptation? Or should it simply reflect relatedness? The issue is particularly pertinent in reference to the classification of the great apes and humans.

Any system of classification of organisms is likely to be hierarchical, a group at any one level being completely subsumed within a group at the level above it, and so on, going from species at the bottom to kingdoms at the top. For instance, humans are classified as the genus *Homo* and species *sapiens*, which is contained within the order Primates, which is part of the class Mammalia, which is within the phylum Chordata (vertebrates), which is subsumed under the kingdom Animalia.

Currently there are three major schools of classification that address the hierarchies of living things: numerical taxonomy, evolutionary (or classical) taxonomy, and cladistics. All three systems have been applied within paleoanthropology, with cladistics becoming increasingly popular.

Numerical taxonomy builds hierarchies on the basis of physical similarity—phenetics—and might therefore be thought of as emphasizing the results of evolution, the adaptations that develop. Cladistics, by contrast, concerns itself only with relatedness—phylogenetics—and is not influenced by the results of adaptation. Evolutionary taxonomy takes the middle ground, tempering strict relatedness with recognition of significant adaptations. (There is in fact a branch of cladistics, called 'transformed cladism', which excludes an explicit basis in evolution, but this does not concern us here.)

If evolution proceeded at regular rates, so that after branching two lineages would diverge steadily in terms of morphological adaptations, then the phenetic pattern would be identical with the phylogenetic pattern. But this generally does not happen: sometimes a new lineage will quickly diverge, accumulating many evolutionary novelties that put a great morphological distance between it and its sister species; sometimes a new lineage will remain almost identical to its sister species over vast periods of time, the morphological distance staying minimal while genetic distance increases. As a consequence of these different tempos of evolution, numerical taxonomy will often yield a different pattern from that produced by cladistic analysis.

The choice of one classification system over the other therefore becomes a matter of philosophy: grouping according to 'overall morphological similarity', which emphasizes adaptation, as against grouping according to relatedness. Which is the

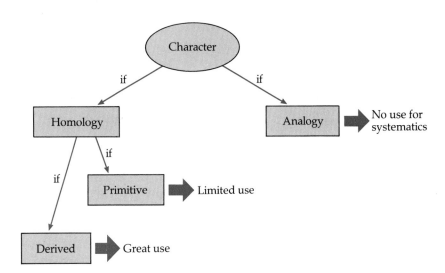

Deducing relationships: A genetic relationship can be deduced between two species only if homologous—not analogous—characters are used. Homologous characters are of two states: primitive and derived. Primitive characters are of limited use in deducing relationships, because they occur in all the species in the group being studied. Derived characters are key to relationships, because they occur in only some of the species under study, and therefore can be used to differentiate within the group.

more 'natural' system? Proponents of numerical taxonomy claim that their analysis is completely objective and completely repeatable, and therefore will reflect meaningful patterns in nature. Cladists argue that the phylogenetic hierarchy is real, whether we discover it or not. There is only one pattern of phylogenetic branching, and that is the path that evolution actually followed: the challenge is being able to infer that pattern from the morphology and other evidence (such as genetics: see unit 9).

A species' morphology is composed of a large suite of anatomical characters: shapes of bones, patterns of muscular attachments, skin color, and so on. Numerical taxonomy proceeds by comparing as wide a range of characters as possible between a group of species, the results of which are known as a multivariate cluster statistics, effectively an average of all the comparisons. The more characters that are included, the more objective the technique is said to be, automatically spitting out a phenetic hierarchy from the assembled cluster statistics. In fact, practitioners frequently have to make choices among several possible patterns, betraying the fact that the method is less objective than is often claimed.

By contrast with numerical taxonomists, biologists who wish to infer evolutionary relatedness among species will not usually use all available characters. The reason is that although many characters that are shared among species are the result of common descent—homology—some will be the result of convergent, or parallel, evolution—analogy. Only homologous characters can be used in reconstructing phylogenies, because these are what link evolutionarily related species together (see unit 4).

The biologist must beware of the trap of convergent evolution, for it is frequently difficult to detect. Convergent evolution can operate on a gross level, for instance in producing virtually identical 'wolves' among placental and marsupial mammal populations that were separated for more than 60 million years—the Tasmanian wolf of Australia compared with the true wolf of Eurasia. Or at a discrete level, such as the details of dental and jaw anatomy, something that is particularly pertinent to paleoanthropologists, because dental and jaw parts are the most frequently recovered fossil specimens of hominoids. Convergence results from the independent evolution to similar function.

Even when characters have been reliably identified as homologous rather than analogous, they are not all equally valuable in inferring evolutionary relatedness. In any group of species under comparison, some homologous characters are said to be primitive, some derived: it is the derived characters that uniquely link species.

Primitive characters are those inherited from the ancestral stock for that group. For instance, baboons, chimpanzees and humans all have nails on the ends of their fingers, but these species are not uniquely linked by this character, because New World monkeys and all prosimians have finger nails too. For baboons, chimpanzees, and humans, the possession of finger nails is therefore a primitive character. However, there is a list of some dozen or so characters that are found uniquely among baboons, chimpanzees and humans, and are absent from New World monkeys and prosimians: these represent derived characters for the Catarrhini, the infraorder that encompasses the Old World monkeys, apes, and humans as a group.

Obviously, the classification of homologous characters into primitive and derived is always relative to the level of the hierarchy being considered. For instance, although the possession of finger nails is a primitive character within the Catarrhini, it is a derived character for primates as a whole: it distinguishes them from other mammals. Generally, what are derived characters at one level will become primitive at the next level up (in the species to kingdom direction). Deciding whether a character state is primitive or derived in a particular species comparison is said to be deciding its *polarity*.

So, in order to infer a unique phylogenetic relationship among a group of species, one has to identify derived characters, the evolutionary novelties that separate the species from their common ancestor. This, simply stated, is the principle of cladistics: a collection of species with shared derived characters that emerged from a single ancestral species is said to be a monophyletic group, or clade; and a diagram indicating relationships is a cladogram.

The cladistic approach was originally developed by the German systematist Willi Hennig in 1950,

and in recent years it has become the approach of choice for many researchers in paleoanthropology. As a result the literature is becoming littered with cladistic analyses and cladistic terminology, which unfortunately includes real tongue twisters. For instance, shared derived characters are *synapomorphies*. Shared primitive characters are *symplesiomorphies*. A derived character not shared with other species is an *autapomorphy*. And convergent characters are *homoplasies*.

There are therefore two steps in determining relationships between species. First, homologies must be separated from homoplasies, which requires careful attention to the traps of functional convergence. Second, polarities of homologous character states must be decided upon: are they primitive (plesiomorphic) or derived (apomorphic). How is polarity determined?

Suppose, for example, one is looking at the bony ridge above the eyes, which is found in chimpanzees, gorillas, and the human lineage, but not in orangutans. Does this mean that this brow ridge is a synapomorphy (shared derived character) linking the three as a clade? Or could it be a symplesiomorphy (shared primitive character) for

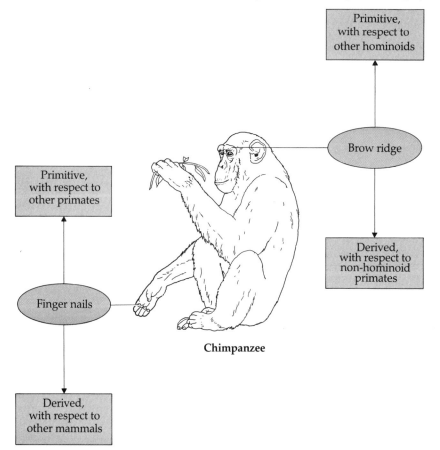

Chimpanzee

Relative status of characters: The state of a character depends on the reference point. For instance, finger nails in an ape are primitive in relation to other primates, because all other primates have finger nails: therefore, finger nails would not serve to distinguish apes from, say, monkeys. However, finger nails are a derived character for primates as a whole, because no other mammals have them: finger nails therefore serve to distinguish primates from other mammals. The second character illustrated here—brow ridges—are found only in hominoids, not in other primates, and are therefore derived for hominoids: they distinguish apes from monkeys. In a chimpanzee, however, brow ridges would be considered as primitive with respect to other hominoids: the character would not distinguish a chimpanzee from, say, a gorilla.

hominoids, which happens to have been lost in the orangutan? The answer is obtained by looking further down the hierarchy, at more distantly related species. This is known as an outgroup comparison. In this case, one would look at a gibbon and also an Old World monkey, for instance. The brow ridge happens to be absent in Old World monkeys, which implies that indeed it is a synapomorphy for the African apes and humans, which by these criteria therefore form a monophyletic group, or clade.

No one, however, likes to base such a judgment on one character. Most analyses therefore include the survey of many characters. The importance of multicharacter comparison is revealed by the fact that one often finds that one subset of characters might imply one pattern of relationship while a second subset points to another. Cladistic analysis of hominoids is no exception. The conclusion from this apparent confusion is that anatomical characters are often extremely difficult to assess and interpret.

For instance, one cladistic analysis of the hominoids in recent years ranked chimpanzees, gorillas, and orangutans as a monophyletic group, leaving humans as a separate clade. A second analysis showed humans and orangutans as a clade, with chimpanzees and gorillas as a second clade. However, most cladistic analyses favor chimpanzees, gorillas, and humans as a monophyletic group, with the orangutan separate, although the preference is not particularly strong.

Now, just suppose that this phylogenetic pattern is correct—and there is support for it from molecular data (see unit 9)—surely it should be reflected in the formal classification? Traditionally, humans and their direct ancestors have been assigned to the family Hominidae, while the African apes and the orangutan occupy a separate family, the Pongidae. Such a grouping certainly reflects 'overall morphological similarity', because humans have diverged dramatically from the apes, but it ignores strict phylogeny.

If phylogeny is to be accurately reflected in classification, then one possibility, as suggested recently by Lawrence Martin of the State University of New York at Stony Brook is as follows. Hominidae would include the African apes in addition to humans, with orangutans occupying the family Pongidae. However, Martin prefers a second possibility, which reflects even more strongly the close phylogenetic relationship between humans and African apes. In

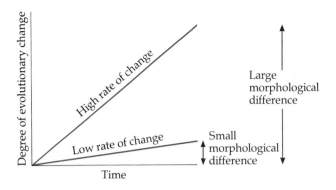

Evolution and morphological change: When two species diverge they may become morphologically quite distinct if evolutionary change is rapid: the difference between humans and African apes, for instance. Or they may remain very similar, if change is slow: compare the chimpanzee and the gorilla.

this alternative, the family Hominidae would encompass humans, the African apes and orangutans, with subfamily status being given to appropriate groups within it. (For comparison, there are 13 genera of Old World monkeys in the family Cercopithecidae.)

Accurate in cladistic terms this may be, but pheneticists and evolutionary taxonomists demur. Classification should also reflect the very drastic ecological shift that has occurred in the hominid line compared with its ape cousins, they contend. Maintaining family status for the apes on the one hand with separate family status for humans on the other is therefore appropriate. Indeed, as mentioned in unit 1, Julian Huxley suggested in 1958 that, because our intellectual and psychological evolution so distances us from the rest of the animal

> **Kingdom** Animalia
> **Phylum** Chordata
> **Class** Mammalia
> **Order** Primates
> **Family** Hominidae
> **Genus** Homo
> **Species** Sapiens

Classification of humans: Traditionally, humans have been classified as the only species in the genus *Homo*, which was the sole occupant of the family Hominidae. This complete separation from the African apes, which were in the family Pongidae in company with the orangutan, was in recognition of the very different adaptions between humans and apes. But, see figure on next page.

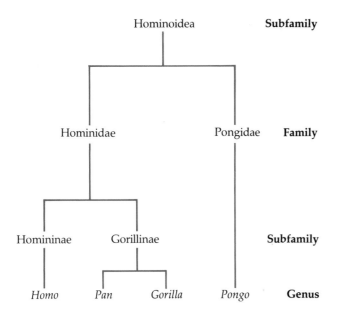

New views of classification: A classification that reflects strictly the evolutionary relationships among the hominoids puts humans and the African apes in the same family — Hominidae — with the orangutan the sole occupant of the family Pongidae. Humans and African apes are then separated into subfamilies.

world, humans might be thought to deserve an entire kingdom of their own: Psychzoa.

If the different philosophies of classification seem routinely to provoke debate, then the business of assigning the appropriate 'rank'—genus, subfamily, family, and so on—among hominoids appears guaranteed to stoke it yet further.

Key questions:

• Why do phenetic and phylogenetic patterns usually differ?
• Which is the more 'natural' system: numerical taxonomy or cladistics?
• Why are derived characters the key to reconstructing phylogenies?
• How important is 'ranking' in describing patterns of classification?

Key references:

Peter Andrews, 'Family group systematics among Catarrhine primates', in *Ancestors: the hard evidence*, edited by Eric Delson, Alan R. Liss Inc., 1985.

Lawrence Martin, 'Relationships among extant and extinct great apes and humans', in *Major topics in primate and human evolution*, edited by B. Wood, L. Martin, and P. Andrews, Academic Press, 1986, pp 161–187.

M.M. Miyamoto *et al.*, 'Molecular systematics of higher primates', *Proceedings of the National Academy of Sciences of the United States of America*, vol 85, p 7627 (1988).

Mark Ridley, *Evolution and classification*, Longman, 1986.

Ian Tattersall and Niles Eldredge, 'Facts, theory and fantasy in human paleontology', *American Scientist*, vol 65, pp 204–211 (1977).

7 / Science of burial

The fossil and archeological records are the principal sources of evidence upon which human prehistory is reconstructed. Unless that evidence can be interpreted with some confidence, the reconstruction, however convincing, may not be valid. During recent years there has developed a tremendous emphasis on understanding the multifarious processes that impinge on bones and stone artifacts that become part of the record. The science of taphonomy, as this pursuit is known, has revealed that the prehistoric record is littered with snares and traps for the unwary.

A taphonomist in a pessimistic mood has been heard to argue that, because of the countless complicating factors that can plant false clues in the record, the chances of reconstructing the past are virtually nil. There is more generally, however, a sense of optimism that step by step specific problems in taphonomy are being solved. Through a combination of ever more careful study of material from the prehistoric record and the development of ingenious experiments and observations on modern material, it is becoming possible to scrutinize the material evidence of human history with the required degree of confidence.

Death is a bewildering, dynamic process in the wild. First, many animals meet their end in the jaws of a predator rather than passing away peacefully in their sleep. Once the primary predator has had its fill, scavengers, which in modern Africa would be hyaenas, jackals, vultures and the like, move in. The carcass is soon stripped of meat and flesh and the softer parts of the skeleton, such as vertebrae and digits, are crushed between powerful jaws. The remaining bones dry rapidly under the sun. Even in this initial phase the skeleton is probably partially disarticulated, hyaenas having torn off limbs and other body parts to be consumed in the crepuscular peace of their dens. Passing herds of grazing animals bring a new phase of disarticulation and disintegration as hundreds of hoofs kick and crush the increasingly fragile bones. Within a few months of a kill the remains of, for example, a zebra, will be scattered over an area of several hundred square metres, and a large proportion of the skeleton will apparently be missing. Some of the skeleton may indeed be miles away, mouldering in a hyaena's den. Some bones will have been shattered and disintegrated into minuscule pieces. Other bones will have been compressed into the ground by the pressure of passing hoofs. Only the toughest skeletal parts, such as the lower jaw and the teeth, remain intact.

As such fate awaits most animals in the wild, it is perhaps not surprising that the fanfared announcements of ancient hominid discoveries typically mean an interesting jaw, or arm bone, or, rarely, a complete cranium has been found. The most complete skeleton of an early hominid unearthed so far is the famous 'Lucy', whose collection of fragmented 3 million-year-old bones represents just 40 per cent of her original self. Nothing else approaches this degree of completeness.

In order to become fossilized a bone must first be buried, preferably in fine alkaline deposits and preferably soon after death. Most hominid fossils have been found near to ancient lakes and rivers, partly because our ancestors, like most mammals, were highly dependent on water; and partly because these provide the best depositional environments where fossil formation is favored.

As it happens, the forces that can bury a bone—for example, layers of silt from a gently flooding river—can later unearth it as the river 'migrates' back and forth across the flood plain through many thousands of years. When this occurs the bones are subject once again to sorting forces: light bones will be transported some distance by the river, perhaps to be dumped where flow is slowed, while heavier bones are shifted only short distances. Anna K. Behrensmeyer, a leading taphonomist, identifies transport and sorting by moving water as one of the most important taphonomic influences. Abrasions caused when a bone rolls along the bottom of a river or stream are tell-tale signs of such activity, as are the characteristic size profiles and accumulations in slow velocity areas of an ancient channel. For hominid remains, the result of all this activity is often the accumulation of hundreds of teeth, and little else, as the researchers working along the lower Omo River in Ethiopia know only too well.

Large numbers of hominid fossils have been recovered from the rock-hard breccia of a number of important caves in South Africa. At one time, it

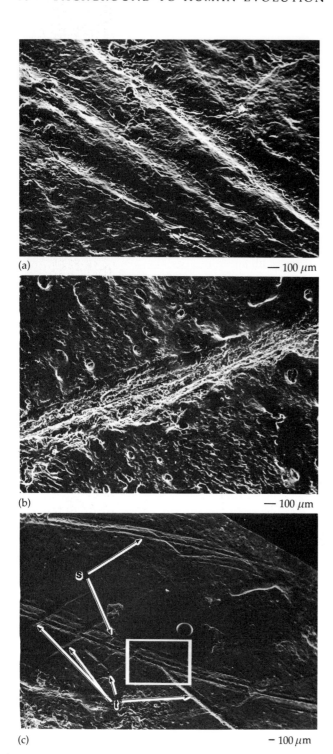

(a) — 100 μm

(b) — 100 μm

(c) — 100 μm

Bone surfaces under the electron microscope:
(a) The surface shows the round-bottomed groove made by a hyaena gnawing at a modern bone.
(b) A sharp stone flake makes a V-shaped groove in a bone surface (modern).
(c) This fossil bone from the Olduvai Gorge carries carnivore tooth marks (t) and stone flake grooves (left at center): the scavenger activity followed the hominid's on this occasion. (Courtesy of Pat Shipman and Richard Potts.)

was thought that the hominids lived in the caves and that the bones of other animals found with them were remains of food brought there to be consumed in safety. In addition, the fractures and holes present in virtually all the hominid remains were considered to be the outcome of hominid setting upon hominid with violent intent. In many ways, the South African caves represent one of the most severe taphonomic problems possible, but with years of patience a group of workers (and in particular C.K. Brain) has cut through the first impressions and progressed a little closer to the truth.

Most of the bone assemblages were almost certainly the ramains of carnivore meals accumulated over very long periods of time. The profile of skeletal parts present matches what would be expected after carnivores had eaten the softer parts. And the damage recorded in the hominid crania was simply the result of rocks and bones compressed into them as the cave deposits mounted. Exactly how much time is represented in these fascinating accumulations is difficult to determine. But the question, as in many taphonomic investigations, is a key one.

One area of investigation in which taphonomic analysis has been particularly crucial in recent years is in the study of ancient assemblies of bones and stones—in other words, putative living sites. Some of the best known and oldest of these occur in the lowest layers of Olduvai Gorge, Tanzania, which are dated to almost 2 million years ago. These concentrations of broken bones and chipped stones have long been assumed to be the product of hunting and gathering activity such as is seen among some technologically primitive peoples today. And the occurrence of such sites appears to increase in frequency through time, giving the impression of an unbroken trail of tell-tale litter connecting people ancient and modern who shared a common lifeway (but see unit 20).

However, in some cases careful taphonomic analysis of the geological setting and the composition of the bone and stone assembly has shown such 'sites' to be the result of water flow, the material having been dumped by a stream in an area of low energy—in other words, the assembly is not an archeological site, but a hydrological jumble. Even when a collection of bones and stones can be shown not to be the result of water flow, there still remains the task of deciding how the various materials reached the site, and whether they were related: specifically, did early hominids use the stones to butcher carcasses?

Taphonomists have determined the stages through which bones go as they lie exposed to the elements—this is the process known as weathering, which can be calibrated. By looking at the degree of weathering evident in a fossil bone, it is therefore possible to determine how long the bone lay on the surface before becoming buried. Applying this technique to the sites at Olduvai, it turns out that in many cases bones accumulated over periods of 5 to 10 years, which would be unheard of in modern hunter–gatherer sites, which are occupied only briefly.

Nevertheless, in the late 1970s and early 1980s, several researchers discovered on the surface of a small percentage of the Olduvai bones what appeared to be marks made by stone tools. So, although the sites might not have been typical hunter–gatherer home bases, there did seem to be a connection between the bones and the stones: the hominids almost certainly were eating meat. By looking at the pattern of distribution of cut marks over a bone—on the shaft as against the articular ends, for example—investigators can get some idea of whether the marks were made during the disarticulation of a carcass or simply the removal of meat or skin from the bone.

Determining the identity of marks on the surface of fossil bones is an important taphonomic activity: gnawing carnivores and nibbling porcupines can all leave their signatures. So too, it seems, can sand grains. During 1986 Behrensmeyer and two colleagues from the Smithsonian Institution, Washington DC reported that bones trampled in sandy sediment can sustain abrasions that are virtually indistinguishable from genuine stone-tool cut-marks. 'Microscopic features of individual marks alone provide insufficient evidence for tool use versus trampling', warn Behrensmeyer and her colleagues. 'If such evidence is combined with criteria based on context, pattern of multiple marks and placement on bones, however, it should be possible to distinguish the two processes in at least some cases bearing on early human behavior.'

Taphonomy adds yet another caveat to the unwary reading of the prehistoric record.

Key questions:

• What is implied by the fact that the great majority of hominid fossil remains have been recovered from sediments layed down near sources of water, such as streams and lakes?
• Why is the fossil record of the African great apes virtually non-existent for the past 5 million years—during which time the hominid record is relatively good?
• Fossil fragments from almost 500 hominid individuals representing perhaps four species over a period of 4 million years ago to 1 million years ago have been recovered from the Lake Turkana region of Kenya: what percentage does this represent of the original populations?
• What is the single most important factor in shaping the life history of a fossil?

Key references:

Anna K. Behrensmeyer, 'Taphonomy and the fossil record', *American Scientist*, vol 72, pp 558–566 (1984).
Anna K. Behrensmeyer and Andrew P. Hill, *Fossils in the making*, University of Chicago Press, 1980.
Anna K. Behrensmeyer *et al.*, 'Trampling as a cause of bone surface damage and pseudo-cutmarks', *Nature*, vol 319, pp 768–771 (1986).
Pat Shipman, *Life history of a fossil*, Harvard University Press, 1981.

8 / Primate heritage

Homo sapiens is one of 185 species of living primate, which collectively constitute the order Primates. (There are 22 living orders in the class Mammalia (mammals), which includes the bats, rodents, carnivores, elephants, and marsupials.) Just as we, as individuals, inherit many resemblances from our parents but also are shaped by our own experiences, so it is with species within an order: each species inherits a set of anatomical and behavioral features that characterize the order as a whole, but each species is also unique, the result of its own evolutionary history.

Matt Cartmill, of Duke University, says of anthropology that: 'Providing a historical account of how and why human beings got to be the way they are is probably the most important service to humanity that our profession can perform.' An understanding of our primate heritage is the starting point for writing that historical account. In this unit we will look at what it is to be a primate, in terms of anatomy and behavior.

The study of primates—primatology—has undergone important changes in recent years, for two reasons. First, ecological research (itself in its scientific infancy) has been thoroughly incorporated into primate studies. As a result, primate biology can be interpreted within a more complete ecological context. Second, the science of sociobiology is facilitating a keener insight into the evolution of social behavior. And primates, if nothing else, are highly social animals. Modern primatology therefore promises to be the focus of some of the most serious intellectual challenges of behavioral ecology.

Modern primates can be divided into four groups: the strepsirhines, which include lemurs, lorises, and bushbabies; New World monkeys, such as the marmosets, spider monkey, and howler monkey; Old World monkeys, such as macaques, baboons, and colobus monkeys; and the hominoids, which

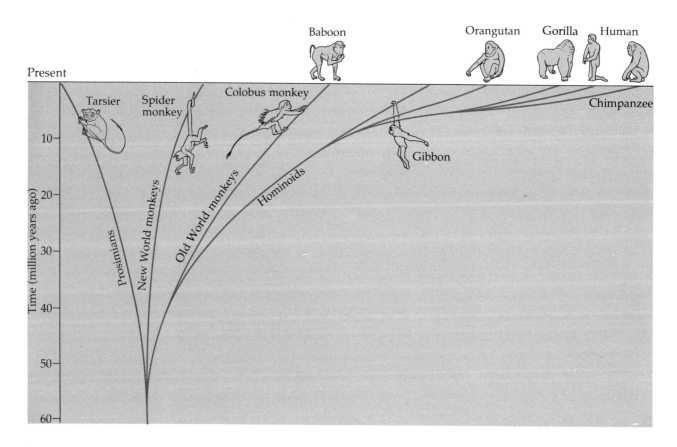

Primate family tree.

comprise apes and humans. The strepsirhines are sometimes known as prosimians, while the monkeys and apes are known collectively as anthropoids. Twenty-eight of the 185 modern primate species live in Madagascar (the lemurs), with about 50 each in Central and South America, Africa, and Asia. There are no native, modern primate species in Europe, North America, or Australia.

Modern primate species vary enormously in size, going from the diminutive mouse lemur, which weighs in at 80 grams, to the gorilla, more than 1000 times its size. But, whatever their size, primates are quintessentially animals of the tropics, and particularly, the tropical rainforest. Although different primate species occupy every major type of tropical environment—from rainforest, to woodland, shrubland, savannah, and semidesert scrub—80 per cent of them are creatures of the rainforest. Several Old World monkeys and one ape—the mountain gorilla—live in temperate and even subalpine zones. Among primates *Homo sapiens* is unique in ranging so wide geographically and in tolerating so extreme a variety of environments.

Although humans have so clearly departed from their primate roots in colonizing so broad a range of habitats, many of the characteristics that we often think of as separating us from other primates—such as upright walking, great intelligence, and extreme sociality—are in fact extensions of, rather than discontinuities with, what it is to be a primate. We should therefore ask, what is it to be a primate?

Surprisingly, this question, which essentially asks for a definition of 'primate', has proved rather difficult to answer concisely. 'It has, in fact, been a common theme throughout the literature on primate evolution that primates lack any clear-cut diagnostic features of the kind found in other placental mammals', notes Robert Martin, of the Anthropological Institute, Zurich. The difficulty, he suggests, stems from an overemphasis on 'skeletal features identifiable in the fossil record'. If, instead, one looks at living primate species, encompassing all aspects of their anatomy and behavior, a definition constructed from universal or near-universal characteristics is possible, says Martin.

'Primates are typically arboreal inhabitants of tropical and sub-tropical forest ecosystems', begins Martin's definition. It goes on to describe features of hand and foot anatomy, overall style of locomotion, visual abilities, intelligence, aspects of reproductive anatomy, life-history factors (such as longevity and reproductive strategy), and dental architecture. The definition, overall, is of species that have a rather special niche in the world. As Rockefeller University anthropologist Alison Jolly recently put it: 'If there is an essence of being a primate, it is the progressive evolution of intelligence as a way of life.'

Some of the key components of the definition are as follows:

Primate hands and feet have the ability to grasp, and are therefore equipped with opposable thumbs and opposable great toes. Humans are an exception here, the foot having lost its grasping function in favor of forming a 'platform' adapted to habitual upright walking. Fingers and toes have nails, not claws; and finger and toe pads are broad and ridged, which aids in preventing slippage on arboreal supports and in enhancing touch sensitivity.

Primate locomotion is hindlimb-dominated, whether it is vertical clinging and leaping (the small species), quadrupedal walking (monkeys and the African great apes), or brachiation (apes). In each case the center of gravity of the body is located near the hindlimbs, which produces the typical diagonal gait (forefoot preceding hindfoot on each

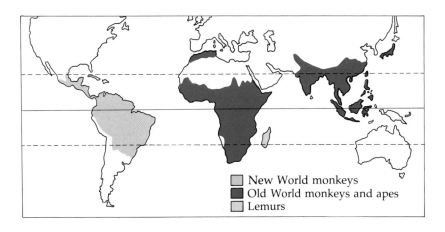

Geographical distribution of living primates.

New World monkeys
Old World monkeys and apes
Lemurs

Modes of primate locomotion: The monkey (top right) walks quadrupedally, while the gibbon (top left) is an adept brachiator (it swings from branch to branch like a pendulum). The orangutan (mid-left) is also adept in the trees, but as a four-handed climber. The gorilla (bottom left), like the chimpanzee, is a knuckle walker (it supports the weight through the forelimbs on the knuckles of the hand rather than using a flat hand as the monkey does). The tarsier (foreground) moves by vertical clinging and leaping. The hominid (right) is a fully-committed biped. Note also, the grasping hands and forward-pointing eyes characteristic of primates. (Courtesy of John Gurche/Maitland Edey.)

side). It also means that the body is frequently held in a relatively vertical position, making the transition to habitual bipedalism in humans less of a dramatic anatomical shift than is often imagined.

Vision is greatly emphasized in primates, while the olfactory (smell) sense is diminished. In all primates the two eyes have come to the front of the head, producing stereoscopic vision, to a greater extent than in other mammals. Some primates (the diurnal species) have color vision, but this does not discriminate the order from many other vertebrate groups. The swinging of the eyes from the side of the head to the front, combined with the diminution of olfaction, produces a shorter snout; but this is also in conjunction with a reduction in the number of incisor and premolar teeth from the ancestral condition of 3 incisors, 1 canine, 4 premolars, and 3 molars (denoted, 3.1.4.3) to a maximum of 2.1.3.3. (Strepsirhines and New World monkeys have this latter pattern, whereas Old World monkeys and hominoids have one fewer premolar.)

Partly because of the emphasis on vision, primate brains are bigger than in other mammalian orders; but the increase also reflects a greater 'intelligence'; the lemurs, lorises and the rest of the strepsirhines are, however, less well endowed than monkeys and apes. Tied to this enhanced encephalization is a shift in a series of life history factors: animals with large brains for their body size tend to have a greater longevity and a low potential reproductive output. For instance, gestation is long relative to maternal body size, litters are small (usually one), and offspring precocious; age at first reproduction is late, and interbirth interval is long. 'Primates are, in short, adapted for slow reproductive turnover', observes Martin.

If we think of ourselves as animals, of what

physically and behaviorally we habitually do, this discursive definition describes us too, apart from the fact that we do not live in trees. For instance, a quarterback would not be able to stand behind his offensive line and accurately throw a deep pass, unless he were a primate. Hindlimb-dominated locomotion, grasping, touch-sensitive hands, stereoscopic vision, and intelligence—all are required here, and all are general characteristics of primates. More historically, when hominids first started making stone tools, they were not 'inventing culture' in the sense that is often used, but merely applying primate manipulative skills to a new task. And, although it is true that even by primate standards *Homo sapiens* is particularly well endowed mentally, our generous encephalization is an extension of just another primate trait.

Later we will be returning to some of these and other themes, particularly the issue of life-history strategy and brain size (units 10 and 27). Here, we will address the question of how primates got to be the way they are, of how a small, ancestral mammal species developed the above suite of characteristics.

The first systematic attempt to account for the differences between primates and other mammals was made early in the twentieth century by British anatomists Grafton Elliot Smith and Frederick Wood Jones. Ancestral primates, and, by extrapolation, humans, were different from other mammals, they argued, as a result of adaptation to life in the trees. Grasping hands and feet provided a superior mode of locomotion, they said, while vision was a more acute sensory system than olfaction in among the leaves and branches.

But, as Cartmill noted, 'The arboreal theory was open to the most obvious objection that most arboreal mammals—opossums, tree shrews, palm civets, squirrels, and so on—lack the short face, close-set eyes, reduced olfactory apparatus, and large brains that arboreal life supposedly favored'.

The British anthropologists valiantly defended their theory, invoking ingenious and often inconsistent lines of argument. In any case, the arboreal theory was modified and extended in the 1950s by another British researcher, the eminent Sir Wilfrid Le Gros Clark, and it thrived for another two decades, until Cartmill felled it in 1972.

In reassessing the arboreal theory in the early 1970s, Cartmill applied biologists' most powerful tool: comparative analysis. 'If progressive adaptation to living in trees transformed a treeshrew-like ancestor into a higher primate, then primate-like traits must be better adapted to arboreal locomotion and foraging than their antecedents', reasoned Cartmill. In other words, if primates are truly the ultimate in adaptation to arboreal life, you would expect that they would be more skillful aloft than other arboreal creatures. 'This expectation is not borne out by studies of arboreal nonprimates.' Squirrels, for instance, do exceedingly well, with divergent eyes, a long snout, and no grasping hands and feet, often displaying superior arboreal skills to those of primates. 'Clearly, successful arboreal existence is possible without primate-like adaptations', concluded Cartmill.

If the close-set eyes and grasping hands and feet were an adaptation to something other than arboreality, what was it? Once again Cartmill used the comparative approach to find an answer. And that answer is embodied in the Visual Predation Hypothesis. Boldly put, the hypothesis states that the suite of primate characteristics is an adaptation by a small arboreal mammal to stalking insect prey, which are captured in the hands.

Cartmill looked for individual elements of the primate suite in a range of other species. For instance, chameleons have grasping hindfeet, which they use to steady themselves when approaching insect prey on slender branches. Some South American opossums do something similar, capturing their prey by hand or mouth. And, of course, the convergence of the eyes is to be seen in many predatory animals that need to be able accurately to judge distance, such as cats, owls, and hawks.

'Most of the distinctive primate characteristics can thus be explained as convergence with chameleons and small bush-dwelling marsupials (in the hands and feet) or with cats (in the visual apparatus)', concluded Cartmill. 'This implies that the last common ancestor of the extant primates... subsisted to an important extent on insects and other prey, which were visually located and manually captured in the insect-rich canopy and undergrowth of tropical forest.' In this respect, the last common ancestor would have subsisted much as modern tarsiers, the mouse lemur, and some lorises do today. These species should not be thought of as 'living fossils', because, like humans, they too are the products of 60 million years of evolution. It is simply that their ecological niche is like their ancestors' was.

Among living primates there is no single 'primate diet': insects, gums, fruit, leaves, eggs, and even other primates—all are to be found on the menu of one primate species or another, and most species regularly consume items from two or more of these categories. The key factor that determines what any individual species will principally subsist on is body size. Small species have high energy requirements per unit of body weight (because of a high relative metabolic rate), and they therefore require food in small, rich packets. Leaves, for instance, are simply too bulky and require too much digestive processing to be of value to small primates. Because of their reduced relative energy demands, large species have the luxury of being able to subsist on bulky, low quality resources, which are usually more abundant. From the small to the large species, the preferred foods shift, roughly speaking, from insects and gums, to fruit, to leaves.

Upon this basic equation there is, however, a good deal of variation. As Richard points out, 'Almost all primates, regardless of size, meet part of their energy requirements with fruit, which provides a ready source of simple sugars'. What sets the basic equation, she says, is 'how they make up the difference in energy and how they meet their protein requirements'. This is where body size is crucial, and why, for instance, the bushbaby's staple is insects and the gorilla's is leaves.

Some kind of species ancestral to the primates lived through the mass extinction 65 million years ago that spelled the end of the Age of Reptiles, the dinosaurs being the most notorious. Soon into the subsequent Age of Mammals, 'primates of modern aspect' appeared, about 50 million years ago, beginning an adaptive radiation that included an increase in range of body size, and a concomitant broadening of diet. The 185 modern species are what currently exists of that adaptive radiation, which, in total, probably gave rise to some 6000 species. The known fossil record gives just the briefest of glimpses of this radiation, a sketchy outline at best: somewhere between 60 and 180 fossil primate species can be recognized.

One very striking feature of this evolutionary pattern is that whereas modern monkey species outnumber apes about 10 to 1, 20 million years ago the position was precisely reversed: the Miocene was the Age of the Ape.

Key questions:

• What general trends did the primate order follow through evolutionary time that are common in other mammalian orders?
• What is to be learned from the pattern of the modern Old World anthropoids, in which monkeys outnumber apes about 10 to 1?
• What key adaptations do humans share with nonhuman primates?
• How great a departure is bipedalism from the mode of locomotion of monkeys and apes?

Key references:

Matt Cartmill, 'Basic primatology and prosimian evolution', in *A history of American physical anthropology*, edited by Frank Spencer, Academic Press, 1982.
Alison Jolly, 'The evolution of primate behavior', in *American Scientist*, vol 73, pp 230–239 (1985).
Robert D. Martin, 'Primates: a definition', in *Major topics in primate and human evolution*, edited by Bernard Wood, Lawrence Martin, and Peter Andrews, Academic Press, 1986.
Alison F. Richard, *Primates in nature*, W.H. Freeman, 1985.

9 / Molecular perspectives

The evolutionary history of every living species is encrypted in its genes. In principle, therefore, it should be possible to reconstruct phylogenies—family trees—by gaining access in some direct or indirect manner to that genetic information, and comparing it among related species. Molecular anthropology—as the application of this approach to questions of human origins is known—is making important contributions in two respects. First, in providing information about the *shape* of the hominoid tree, which indicates relatedness. And second, in showing *how long the branches are*, which indicates the length of time lineages diverged from each other. This second aspect embodies what is known as the molecular clock.

The interplay between molecular evidence of relatedness and anatomical evidence, which anthropologists have traditionally used to reconstruct phylogenies, has frequently sparked lively debate, and sometimes downright controversy. The reason is that the inferences derived from these independent lines of evidence have often been somewhat contradictory, both in terms of the shape of the tree and its overall timetable. In recent years a new mood of cooperation has developed, in which researchers are trying to see what these disparities might imply about the nature of their different types of evidence.

Emile Zuckerkandl and Linus Pauling in 1962 were first to suggest that molecular biological evidence—specifically from proteins—might be a powerful tool for building family trees. The proposal was based on the assumption that once a lineage had split into two separate taxa, genetic mutations would accumulate independently and at a steady rate in the genes of the two new lines: the longer they were separated, the more different their genes would become.

Zuckerkandl and Pauling realized that not all genes accumulate mutations at the same rate, because of differences in constraints of natural

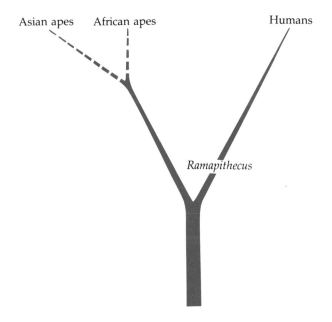

Premolecular evidence: Asian and African apes thought to be closely related; hominids diverged from apes prior to 15 million years ago, because of position of *Ramapithecus* as putative early hominid.

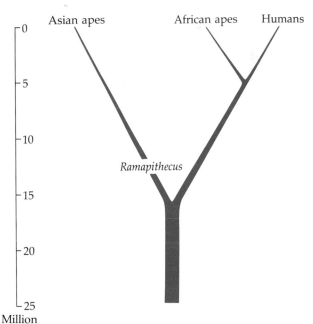

Postmolecular evidence: Asian apes distant from African apes, which diverged from hominids very recently (perhaps five million years ago). *Ramapithecus* cannot, therefore, be a hominid because it lived prior to the African ape/hominid divergence.

A scheme showing envisaged branching patterns: Based on premolecular evidence and postmolecular evidence.

selection: the molecular clock for each gene ticks at a different rate, but ticks regularly nevertheless. Or so it was thought. Whether the molecular clock really is metronomic—in a stochastic sense, that is—or instead varies within and among lineages is a matter of current debate among researchers. This uncertainty makes it imperative that any potential molecular clock data are subject to what is known as the 'rate test' to ensure that legitimate comparisons can be made. If the data pass the test, then the shape—and possible the timetable—of the phylogenetic tree can be inferred.

Morris Goodman of Wayne State University pioneered modern molecular anthropology, and in 1963 published results of immunological comparisons of certain blood proteins from apes and humans. At the time anthropologists considered that the three great apes—chimpanzee, gorilla and orangutan—formed a monophyletic group, with humans being genetically more distant. This picture had developed in the 1940s and 1950s and ran counter to Darwin's and Huxley's earlier conclusions that humans and the African apes were closely related, with the Asian great ape being separate. However, when Goodman published his immunological comparisons, they clearly favored Darwin's and Huxley's interpretations, not the current anthropological wisdom. This was the first clash between molecular and anthropological protagonists.

The next came 4 years later, in 1967, when Allan Wilson and Vincent Sarich of the University of California, Berkeley, produced some dates for Goodman's phylogenetic tree. Using measures of genetic distance by immunological comparisons of blood proteins, Wilson and Sarich said that the human and African ape lineages diverged about 5 million years ago. The figure was arrived at by comparing the genetic distance of the two species in question with the distance from a related group whose origin is known in the geologic record. In this case, the calibration group was Old World monkeys, whose origin was put at 30 million years ago: the genetic distance between humans and African apes was about one-sixth that of the distance between humans and Old World monkey, which translates to 5 million years. This principle of calibration from the fossil record is utilized in all molecular clock calculations.

Anthropologists at the time believed that the human/African ape split was somewhere between 15 and 30 million years ago, based on the putative hominid status of *Ramapithecus*, fossils of which

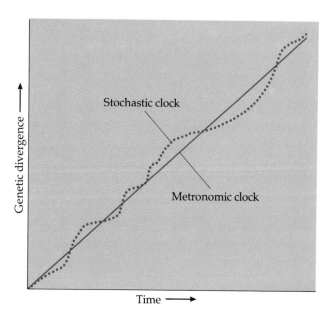

The molecular clock: If genetic mutation were to occur at a constant rate, then biologists would have access to a completely reliable, 'metronomic' molecular clock. In fact, the rate of mutation for any particular region of DNA is likely to fluctuate through time, giving a 'stochastic' molecular clock. By bringing together data on genetic divergence from different regions of DNA it is possible in principle to average out these fluctuations, giving a good, average clock. Because the technique of DNA hybridization effectively compares the entire DNA complement of two related species, fluctuations in mutation rate in different parts of the genome are automatically averaged out.

had been recovered from Pakistan, and various other sites in the Old World. Wilson and Sarich's 5 million year date was therefore not taken seriously.

For the next decade and a half more molecular data of various different kinds were produced. In addition to immunological comparisons, these included electrophoresis of proteins, amino acid sequencing of proteins, restriction enzyme mapping and sequencing of mitochondrial DNA and nuclear DNA, and DNA–DNA hybridization. Almost without exception, these new data favored the human/African ape grouping, with the orangutan more distant. And, although exact figures differed between them, all the molecular data placed the human/African ape split at close to 5 million years ago. Nevertheless, throughout this time most anthropologists resisted the conclusions.

The mood changed in the early 1980s, following the discovery and description of two well-preserved *Sivapithecus* fossil faces, one in Turkey, the other in Pakistan, the principle implication of which was that *Ramapithecus* could no longer be regarded as a hominid. With *Ramapithecus* removed as a putative hominid, the anthropological community came to accept that the time of the human/African ape divergence was closer to 5 million than 15 million years.

The shape of the hominoid tree according to the molecular evidence available in the early 1980s was therefore as follows: gibbons split away first, about 20 million years ago; orangutans next, about 15 million years ago; leaving humans, chimpanzees, and gorillas in an unresolved three-way split, close to 5 million years ago. A three-way split of a lineage is biologically unlikely, and in this case it meant that the timing of the different divergences was so tightly bunched that none of the techniques was able to prise it apart with any confidence.

Meanwhile, most morphologists had since the 1960s accepted the notion of a human/African ape clade, with an African ape clade existing within that. The expectation among molecular biologists, therefore, was that their data would confirm this pattern, showing that the common ancestor of humans and the African apes diverged to produce the human lineage on the one hand and an African ape lineage on the other, which then subsequently split to produce gorillas and chimpanzees.

It was therefore something of a surprise when, in 1984, Charles Sibley and Jon Ahlquist, then of

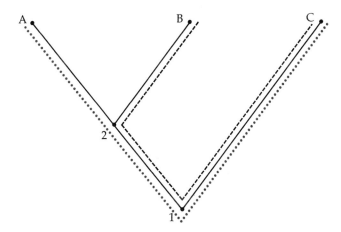

The rate test: The diagram represents two evolutionary events. At 1, a split occurred, leading to species C on one hand and a second lineage on the other. This second lineage split at node 2, leading to species A and B. The rate test says that if the average rate of genetic divergence is the same in all lineages, then the genetic distance from species A to species C (dotted line) should be the same as the genetic distance from species B to species C (dashed line). If, however, gene mutation slowed down in lineage B, then the B to C genetic distance would be shorter than A to C.

Yale University, published data on DNA–DNA hybridization that strongly implied that chimpanzees are more closely related to humans than they are to gorillas. Gorillas evolved from the human/African ape common ancestor between 8 and 10 million years ago, they concluded, leaving humans and chimpanzees briefly sharing a common ancestor of their own, and splitting at between 6.3 and 7.7 million years ago.

The technique of DNA–DNA hybridization effectively compares the entire genome of one species against the genome of another. Any differences between the two will result in reduced interaction of the DNA strands under experimental conditions, and this can be measured and translated into genetic distance. The potential power of the technique is in the averaging of very large numbers: the human genome, minus multiple repeat sequences which are removed in the experiment, amounts to 2 billion individual bases. Although the technique cannot reveal what exactly has changed in the DNA of two species under comparison, in principle it is capable of reflecting virtually every base change that has taken place. How reliable the technique is in practice, however, is questioned by some researchers, as is the quality of some of Sibley and Ahlquist's data.

Questions about the reliability of DNA–DNA hybridization notwithstanding, the dozen or so publications of various kinds of molecular data that followed in the next 4 years supported the human/chimpanzee association by a majority of more than 2 to 1. Sibley and Ahlquist tripled the size of their DNA–DNA hybridization dataset, and in 1987 came down even more strongly in support of their original conclusion, albeit with slightly different dates. And since that time Jeffrey Powell of Yale University has confirmed this result, also using DNA hybridization.

Perhaps the most important set of molecular data, published by Goodman and his colleagues in 1987, is the DNA sequence of a 7100-base long region of the β-globin gene locus in hominoids. This was the longest run of sequences available for comparison, and according to some theoreticians significantly exceeds the number required for a reliable conclusion of phylogeny. Sequence data allow one to see exactly what has changed in the DNA, and a phylogenetic tree can be constructed by working out what might be thought of as the shortest route back to the ancestral sequence. Such a technique is known as parsimony analysis, and for Goodman and his colleagues it favored a human/chimpanzee association over a chimpanzee/gorilla association by about 3 to 1.

The anthropologists, meanwhile, had been further scrutinizing anatomical evidence on the hominoid relationship. One researcher concluded, as in the 1940s and 1950s, that the three great apes formed a monophyletic group, excluding humans from an especially close relationship with chimpanzees and gorillas. Another suggested that humans were most closely related to orangutans, with the African apes separate. However, most anthropologists favored the idea of human/African ape clade, but with chimpanzees and gorillas being most closely related to each other. For instance, in a cladistic analysis of hominoids, Lawrence Martin of the State University of New York at Stony Brook concluded that two major character complexes uniquely link the two African apes. These are the knuckle-walking adaptation and the mode of development and structure of molar tooth enamel.

Peter Andrews of the British Museum (Natural History) recently joined with Martin in another cladistic analysis, this time of both the morphological and molecular evidence. For the cladistic analysis of molecular evidence, Andrews and Martin of course had to restrict themselves to character data: specifically, amino acid sequences of proteins and DNA sequences. Data from DNA–DNA hybridization, immunological comparisons, and protein electrophoresis, for instance, cannot be analyzed cladistically, because it is not known exactly what is changing between the various molecules: these techniques give measures only of genetic distance, not specific characters whose polarity can be judged (see unit 6).

Andrews's and Martin's conclusions are intriguing and puzzling. For instance, although cladistic analysis of morphological data link chimpanzees and gorillas together strongly, support for a human/African ape clade is surprisingly weak. By contrast, molecular evidence strongly suggests a human/African ape clade, with humans and chimpanzees being linked together within it. One set of data—molecular or anatomical—must be being incorrectly interpreted in some way. Andrews and Martin note that in the molecular data, whatever shape of human/chimpanzee/gorilla tree is selected, there appears to be a surprisingly high degree of parallel evolution (homoplasy), much higher than would be predicted on simple statistical grounds. Its significance remains obscure for the present.

	Ape/human divergence date: (millions of years)	
	Fossils	Molecules
1980s	5–8	5–8
1970s	15	5
1960s	30	5

Converging evidence: During the past three decades the perceived date for the human/ape divergence based on fossil evidence has steadily decreased, until in 1981 it fell into agreement with the molecular evidence.

If the human/chimpanzee association is correct, then there are certain implications regarding the common ancestor of humans and African apes: specifically in its mode of locomotion. The notion that chimpanzees and gorillas might have independently evolved the knuckle-walking habit from a nonknuckle-walking ancestor is unacceptable to many anatomists, simply because it is such a very complex set of anatomical adaptations: for it to have evolved twice is thought to be unlikely.

The alternative—that the common ancestor of humans and the African apes was a knuckle walker, an adaptation that was lost in the hominid line—is equally difficult to accept. Most anatomists who have studied the issue see no vestigial signs of knuckle walking in the anatomy of living or fossil hominids. Yet, if humans really are uniquely linked with chimpanzees as the molecular evidence seems to imply, one of these alternatives must be correct. It would be a tremendous challenge to comparative anatomy to resolve this one. It should be noted, however, that one morphologist—Colin Groves, of the Australian National University—links humans and chimpanzees in his analysis of the anatomical data.

As we saw in unit 6, in most traditional anthropological texts humans and their direct ancestors are formally classified under the family Hominidae while the three great apes are in a separate family, the Pongidae. Goodman challenged this classification in his 1963 paper, saying that because

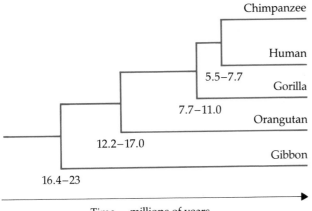

A tree with molecular dates: Data from DNA hybridization give a range of dates for the evolutionary origins of the hominoids. Although many other genetic techniques now give a tree of this shape, none gives as long an interval of time between gorilla divergence and subsequent chimpanzee/human divergence.

humans, chimpanzees, and gorillas formed a natural group, they should all be classified in the same family, the Hominidae. The orangutan would then remain as the only member of the family Pongidae. Given the implications of the molecular data of more recent years, Goodman now suggests that formal classification should reflect an even closer relatedness between humans and African apes. Humans, chimpanzees, and gorillas should be members of the subfamily Homininae; the orangutan would be the sole member of the subfamily Ponginae; these two subfamilies would then make up the family Hominidae.

Similar rationalizations have been suggested by some anthropologists, including Andrews and Martin, though details differ. Most anthropologists,

however, seem reluctant to force humans to share family status—still less, subfamily status—with any other extant species: the resistance reflects attitudes towards the specialness of *Homo sapiens*. The debate comes down to the question of the degree to which classification should reflect genealogy as against adaptation (see unit 6).

Key questions:

● Does the molecular clock tick at different rates in different lineages?
● What is the relative importance of distance measures and character states in molecular data?
● What are the implications of the conflicting conclusions between molecular and morphological data over the human/African ape relationship?
● How should relatedness among the great apes and humans be reflected in formal classification?

Key references:

Peter Andrews and Lawrence Martin, 'Cladistic relationships of extant and fossil hominoid primates', *Journal of Human Evolution*, vol 16, p 101 (1987).

Roy Britten, 'Rates of DNA sequence evolution differ between taxonomic groups', *Science*, vol 231, p 2393 (1986).

John E. Cronin, 'Apes, humans, and molecular clocks: a reappraisal', in *New interpretations of ape and human ancestry*, edited by Russell L. Ciochon and Robert S. Corruccini, Plenum Press, 1983.

Morris Goodman, 'Molecular evidence of the ape subfamily Homininae', in *Evolutionary perspectives and the new genetics*, Alan Liss Inc, 1986, p 121.

C.P. Groves, 'Systematics of the great apes', *Comparative Primate Biology*, vol 1, pp 187–217 (1986).

Special issue on molecular clocks, *Journal of Molecular Evolution*, vol 26, nos 1–2 (1987).

10 / Bodies, brains, and energy

Ideas of human evolution have traditionally been dominated by the supposed intellectual and technological skills displayed by our large-brained ancestors. In fact, as Cambridge University anthropologist Robert Foley has recently stressed, 'Many aspects of human evolution are explicable in terms of the causes and consequences of increased body size.'

This unit will explore the impact of size—of both brains and bodies—on life-history variables and behavioral ecology. We will see why hominids, with their large body size, have many more options open to them in terms of diet, foraging range, sociality, expanded brain capacity, and so on, than, say, the diminutive mouse lemur.

In 1978 Princeton ecologist Henry Horn encapsulated the range of potential ecological options by posing the following set of questions: 'In the game of life an animal stakes its offspring against a more or less capricious environment. The game is won if offspring live to play another round. What is an appropriate tactical strategy for winning this game? How many offspring are needed? At what age should they be born? Should they be born in one large batch or spread out over a long lifespan? Should the offspring in a particular batch be few and tough or many and flimsy? Should parents lavish care on their offspring? Should parents lavish care on themselves to survive and breed again? Should the young grow up as a family, or should they be broadcast over the landscape at an early age to seek their fortunes independently?'

In responding to these challenges the animal kingdom as a whole has come up with a vast spectrum of strategies, ranging from species (oysters, for instance) that produce millions of offspring in a lifetime, upon which no parental care is lavished, to species (such as elephants) that produce just a handful of offspring in a lifetime, each born singly and becoming the object of intense and extensive parental care. In the first case the potential reproductive output of a single individual is enormous, though typically curtailed by environmental attrition; in the second case it is small.

By their nature, mammals are constrained in the range of life-history patterns open to them: mam-

A difference in body sizes: The gorilla and the mouse lemur represent the largest and the smallest of the primates, with the females of the species weighing 93 kilograms and 80 grams respectively. Such differences in body size have many implications for a species' social and behavioral ecology. One of the most dramatic is in potential reproductive output: the female mouse lemur can grow to maturity and, theoretically, leave 10 million descendants in the time it takes a female gorilla to produce just one.

Life-history factors

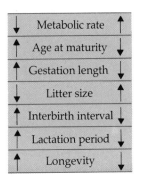

↓ Metabolic rate ↑	
↑ Age at maturity ↓	
↑ Gestation length ↓	
↓ Litter size ↑	
↑ Interbirth interval ↓	
↑ Lactation period ↓	
↑ Longevity ↓	

Life-history factors: Body size affects a large range of life-history factors, as illustrated here. For instance, a large primate will have a long lifespan, will mature late, have a long gestation time and lactation period, will have a long period between litters, but litters will be small (usually one), and basal metabolic rate will be low.

Large body size

Small body size

malian mothers are limited in the number of offspring that can be carried successfully through gestation and suckling. Nevertheless, potential reproductive output can be relatively high if more than one litter is raised each year over a lifetime of several years.

In the order Primates, potential reproductive output is low compared with mammals as a whole, litters being restricted in the vast majority of species to a single offspring. In the parlance of population biology, primates are therefore said to be *K*-selected. (Species with a high potential reproductive output are said to be *r*-selected.) And of all the primates, humans are the most extremely *K*-selected species.

Success in simple Darwinian terms is often measured in the currency of reproductive output, which is determined by a series of interrelated life-history factors. These include age at maturity, length of gestation, litter size, duration of lactation period, interbirth interval, and lifespan.

Some species live 'fast' lives: within a short lifespan they mature early, produce large litters after a short gestation period, and wean early. The result is a large potential reproductive output. Other species live 'slow' lives: within a long lifespan they mature late, produce small litters (a single offspring) after a long gestation period, and wean late. Here, the potential reproductive output is small.

As it happens, the best predictor as to whether a species lives fast or slow is its body size: small species live fast lives, large species live slow lives. As potential reproductive output is highest in species that live fast lives, it might be thought that all species would be small. That some species are large implies that there are benefits in a bigger

body size that trade off for a reduced potential reproductive output.

Such benefits might include (for a carnivore) a different spectrum of prey species or (for a potential prey) better anti-predator defenses. Another potential benefit of increased body size is the ability to subsist on poorer quality food resources. The reason for this is that basal energy demands increase as the 0.75 power of body weight: in other words, as body weight increases, the basal energy requirement *per kilogram of body weight* decreases, a relationship known as the Kleiber curve. This is why mouse lemurs must feed on energy-rich insects and gums, for instance, while gorillas can subsist on energy-poor foliage. A further potential benefit of increased body size is an improved thermoregulatory efficiency. And so on.

The generally close relationship between body size and the value of the various life-history factors is the outcome of certain basic geometric and bioenergetic constraints. The result is that for any particular body size increase there is a more or less predictable change in, say, gestation length, age at maturity, and so on. For each life-history variable, therefore, a log/log plot against body size produces a straight line, with a particular exponent which describes the relationship (0.75 for basal energy needs, 0.37 for interbirth interval (in primates), 0.56 for weaning age, and so on). In effect, such plots *take out* body size in species comparisons.

If basic engineering constraints were all that underpinned life-history factors, then every species would be directly equivalent with every other species *when body weight is taken into account*: all the figures for each life-history variable would fall on the appropriate straight lines. In fact, individual figures often fall above or below the line, indicating a good deal of life-history variation. It is this variation that reveals an individual species' (or, more usually, a group of related species') adaptive strategy.

In recent years researchers have realized that, in addition to body size, brain size is also highly correlated with certain life-history factors, in some cases much more so than is body size.

Among mammals as a whole there is a key dichotomy in developmental strategy that has important implications for life-history measures: this is the altricial/precocial dichotomy. Altricial species produce extremely immature young that are unable to feed or care for themselves. The young of precocial species, on the other hand, are relatively mature and can fend for themselves to a degree.

Life-history factors critically associated with altriciality and precociality include gestation length. In altricial species gestation is short and neonatal brain size is small. Gestation in precocial species is relatively long, and neonatal brain size is large. There is, however, no consistent difference in *adult* brain size between altricial and precocial species. Primates as a group are precocial but *Homo sapiens* is an exception among primates, having developed a secondary altriciality and an unusually large brain (see unit 27).

In addition to the distinction between fast and slow lives according to absolute body size, some species' lives may be fast or slow *for its body size.*

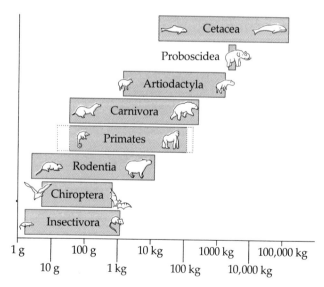

Body size compared: Primates are in the middle range of mammalian body sizes as a whole. Nevertheless, the biology of hominoids is the biology of large mammals. Most mammalian species are concentrated in Rodentia, Chiropter and Insectivora, and are therefore small.

Such deviations have traditionally been explained in terms of classic *r*- and *K*-selection theory. According to this theory, environments that are unstable in terms of food supply (that is, are subject to booms and busts) encourage *r*-selection: fast lives, with high potential reproductive output. Alternatively, stable environments (which are close to carrying capacity and in which competition is therefore keen) favor *K*-selection: slow lives with low potential reproductive output.

As mentioned earlier, primates are close to the *K*-selection end of the spectrum among mammals as a whole, but some primates are less *K*-selected than others. For instance, Caroline Ross has recently shown that, when body size is taken into account, primate species that live in unpredictable environments have higher potential reproductive output than species in more stable environments.

A second factor that influences whether a species might live relatively fast or slow for its body size has recently been identified by Paul Harvey and Daniel Promislow. In a survey of 48 mammal species Promislow and Harvey found that 'those species with higher rates of mortality than expected had shorter gestation lengths, smaller neonates, larger litter, as well as earlier ages at weaning and maturity'. In other words, species that suffer high natural rates of mortality live fast. 'The reason is that species with higher rates of mortality are less likely to survive to the following breeding season and will therefore be selected to pay the higher costs associated with earlier reproduction.'

Again, does the very slow life lived by *Homo sapiens* imply evolution from an ancestor that experienced very low levels of mortality?

Given that most mammals measure less than 32 cm in length, hominids—even the early, small species—must be classified as large mammals. The earliest known hominid species, *Australopithecus afarensis*, stood 1 meter (females) to 1.7 meters (males) tall, and weighed some 30 to 65 kilograms. These general proportions persisted until about 1.5 million years ago with the evolution of *Homo erectus*, which stood close to 1.8 meters (with a much reduced difference between males and females).

Knowing these general body proportions and the estimates of brain size it then becomes possible to make estimates of various life-history factors for the early hominid species, given also what is known of the only extant hominid, *Homo sapiens*. Surely, hominids lived slow lives in the terms of life-history variables, with a vastly increased brain

	r-Selection	K-Selection
Climate	Variable and/or unpredictable; uncertain	Fairly constant and/or predicable; more certain
Mortality	Often catastrophic, non-directed, density independent	More directed, density dependent
Survivorship	High juvenile mortality	More constant mortality
Population size	Variable in time, non-equilibrium; usually well below carrying capacity of the environment; unsaturated communities or portions thereof; ecological vacuums; recolonization each year	Fairly constant in time, equilibrium, at or near carrying capacity of the environment; saturated communities; no recolonization necessary
Intra- and interspecific competition	Variable, often lax	Usually keen
Selection favors	• Rapid development • High maximal rate of increase, r_{max} • Early reproduction • Small body size • Single reproduction • Many small offspring	• Slower development • Greater competitive ability • Delayed reproduction • Larger body size • Repeated reproduction • Fewer larger progeny
Length of life	Short, usually less than 1 year	Longer, usually more than 1 year
Leads to	Productivity	Efficiency

Characteristics of r- and K-selection.

capacity eventually distorting some of them (see unit 27).

In addition it is possible to identify several behavioral ecology traits that would also be associated with large body size, as Foley has done recently. For instance, dietary scope could be broad; day and home ranges could be large; mobility could be high; predator–prey relations would be shifted from that of smaller primates; thermoregulatory efficiency would be improved; sociality would be extended; and enhanced encephalization would be possible.

In sum, studies of life-history strategies have identified body size, brain size, environmental variability, and mortality rates as being crucial to the rate at which a species will live. Much of human evolution may therefore be explicable in terms of a large hominoid exploiting a relatively stable food supply, its stability perhaps being enhanced by virtue of its breadth. Technology may eventually have contributed to this stability by allowing a more efficient exploitation of meat and certain plant foods, thus broadening the diet still further. A reduction in mortality, perhaps through improved anti-predator defense, would further encourage a 'slow' life-history strategy. But the selection pres-

sure leading to increased body size still has to be identified.

Key questions:

• What are the limitations of a simple Darwinian measure of reproductive success?
• At any particular body size, which is the riskier strategy, living fast or living slow?
• Primates as a group are twice as brainy as other mammals: how might this have arisen?
• Could the first hominids have originated in tropical rain forests?

Key references:

Robert Foley, *Another unique species*, Longman Scientific and Technical, 1987.
Paul Harvey, Robert Martin, and Tim Clutton-Brock, 'Life histories in comparative perspective', in *Primate societies*, edited by B.B. Smuts, D.L. Cheney, R.M. Seyfarth, R.W. Wrangham, and T.T. Struthsaker, University of Chicago Press, 1986, pp 181–196.
Robert Martin and Paul Harvey, 'Human bodies of evidence', *Nature*, vol 330, pp 697–698 (1987).
Daniel Promislow and Paul Harvey, 'Living fast and dying young', *American Naturalist* (in press).
Caroline Ross, 'The intrinsic rate of natural increase and reproductive effort in primates', *Journal of Zoology*, vol 214, p 199 (1988).

11 / Bodies, behavior, and social structure

The great majority of primate species are social animals, living in groups that range from two to 200 individuals. Whatever the size of the group, it is the focus of many important biological activities, including foraging for food, raising offspring, and defense against predators. The group is also the center of intense social interaction that has little apparent direct bearing on the practicalities of life: in the human sphere we would call it socializing, the making and breaking of friendships and alliances (see unit 27). The size, composition, and activity of a group defines what is usually known as a species' social organization.

Animal behavior is a far more variable characteristic than, for instance, anatomy or physiology. As a result, an order such as the Primates will display an astonishingly wide range of social organization, in which even closely related species may carry out their daily social lives in very different ways. We saw in unit 10 that body size can have a powerful bearing on many aspects of a species' way of life, but social organization is not one of them. Even if we consider just the apes—

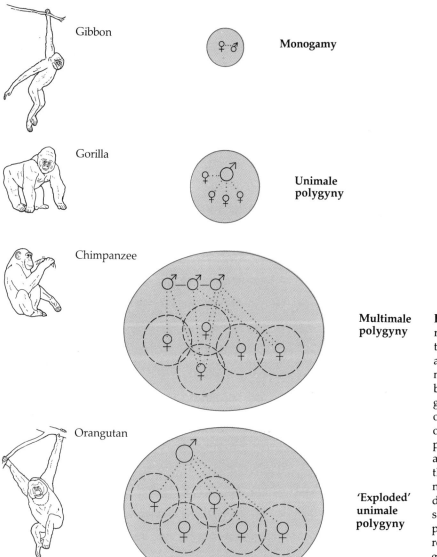

Gibbon — Monogamy

Gorilla — Unimale polygyny

Chimpanzee — Multimale polygyny

Orangutan — 'Exploded' unimale polygyny

Hominoid social organization: The range of social organization among the apes matches that among anthropoids as a whole. Gibbons are monogamous, with no size difference between males and females. In gorillas, a single male has control over a group of females (and their offspring): this is known as unimale polygyny. Single male orangutans also defend a group of females (and their offspring), but the females do not live as a group but instead are distributed over a large area: this is sometimes known as exploded polygyny. In chimpanzees, several related males cooperate to defend a group of widely distributed females (and their offspring): this is an example of multimale polygyny.

the largest of the nonhuman primates—the array of social organization is just about as great as among the primates as a whole.

Highly social creatures ourselves, it may seem odd to ask: 'why should animals live in groups?' But it is in fact a very good biological question, because there are many costs to gregariousness. For instance, a lone individual doesn't have to share its food with another individual, but in a group there is competition for all resources. A lone individual is not exposed to diseases that flourish in communities, which provide a viable host pool for pathogens. A lone individual is much less conspicuous to predators than is a group of individuals. And so on. Clearly, as most primates do live in groups, the benefits must outweigh the costs.

This unit will discuss current thinking about the benefits—causes—of living in groups. It will also examine some of the consequences of group living: not the costs mentioned, but the ways in which individuals might adapt behaviorally and anatomically to different types of social structures.

In order to have a feel for some of the details of social organization, and the range to be found among primates, we will first survey the social lives of the apes: gibbon (and siamang), orangutan, chimpanzee, and gorilla.

Gibbons and siamangs are the smallest of the apes (sometimes called the lesser apes), and live in forests in southeast Asia. The basic social structure of these highly acrobatic, arboreal creatures is very similar, being made up of a monogamous mating pair, plus up to three dependent offspring. Gibbons are territorial, and eat a diet of fruit and leaves. On reaching maturity, the offspring leave the natal group and eventually establish one of their own by pairing with another young adult of the opposite sex. Mature males and females are essentially the same body size. Gibbons are a good example of life-long monogamy.

The other Asian ape, the orangutan, is much larger than the gibbon and pursues a very different lifeway, although it too is highly arboreal. The core of social organization is a single mature female with her offspring. The mother and offspring occupy a fairly well-defined home range, which usually overlaps with that of one or more other mature females and their offspring. Males are rather solitary creatures, occupying a large territory that usually contains the home ranges of several mature females with whom he will mate. Males defend their territories against incursion by other males, and are

♂ : Access to mature females

♀ : Access to food resources

Different reproductive strategies: For a female primate the variable that determines ultimate reproductive success is access to food resources. By contrast, a male's reproductive success is limited by his access to mature females. This difference critically influences the overall social structure of primate societies.

about twice the size of females. The mating system is therefore one of a loosely organized harem, one male mating with several females (technically known as unimale polygyny).

Gorillas, the largest of the apes, live in the forests of central and west Africa. These animals have a similar mating system to that of the orangutan—unimale polygyny—although the ecology and organization is rather different. Predominantly terrestrial animals that live on low-quality herbage that is usually found in abundant but widely dispersed patches, gorillas live in close groups of two to 20 individuals. The adult male—the silverback—has sole mating access to the mature females, whose immature offspring also live in the group. Mature males compete for control of the group. Nevertheless, from time to time a female, usually a young adult, will transfer from one group to another, seemingly as a matter of free choice. New groups are established when a lone silverback begins to attract transferring females. As with orangutans, male gorillas are twice the size of females.

Chimpanzees, terrestrial and arboreal omnivores, live in rather loose communities of between 15 and 80 individuals, a mixture of mature males and mature females and their offspring. Unlike savannah baboons, which live in close, cohesive troops of mature males and females and their offspring, sometimes numbering 200 individuals in all, chimpanzee communities are maintained by occasional contact between males and females. The core of chimpanzee social life is a female with her offspring, often to be found by themselves but sometimes with other females and their offspring. Each female maintains a core area, which usually overlaps with that of one or more other females. By contrast with orangutans, single chimpanzee males are unable to maintain exclusive control of a group of female home ranges. Instead, a group of males defends the community range against the males of

neighboring communities. Mating in chimpanzee communities is promiscuous, with each estrus female copulating with several males. The social organization is therefore known as multimale polygyny.

A key feature of chimpanzee social organization is that, contrary to the general pattern of multimale societies among primates, males remain in their natal group while young adult females transfer (or are sometimes kidnapped) to other communities. The result is that the group of adult males who are cooperating to defend their community are usually rather closely related to each other. Adult males are typically about 25 to 30 per cent larger than females.

Among the apes, then, there is to be found monogamy, unimale polygyny, and multimale polygyny. (Polyandry—one female having exclusive access to several mature males—which is rare in mammals generally, is absent here.) Some of the aspects of group living raised by this spectrum of social organization include: how big will a social group be? What is the ratio of adult females to adult males? Among which sex is there the greater degree of relatedness? What difference is there in the size of males and females?

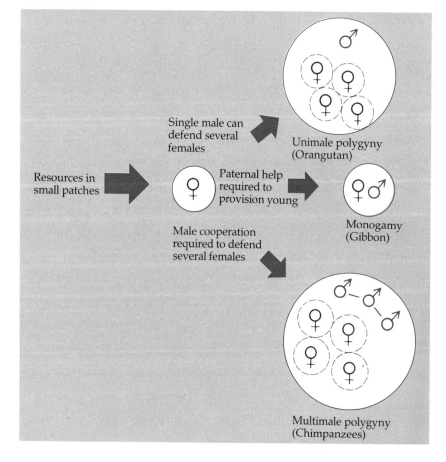

Single male can defend several females

Unimale polygyny (Orangutan)

Resources in small patches

Paternal help required to provision young

Monogamy (Gibbon)

Male cooperation required to defend several females

Multimale polygyny (Chimpanzees)

Distribution, with small resource patches: When food comes in patches too small to support more than one mature individual, females will forage singly (with offspring). If a male can defend a 'community' of lone females, unimale polygyny will result, as with the orangutan. If males can defend only one female, or if paternal help is required in raising offspring, monogamy will result, as with the gibbon. If a community of females can be defended only by several males, then a group of related males will defend a number of unrelated females, as in chimpanzees.

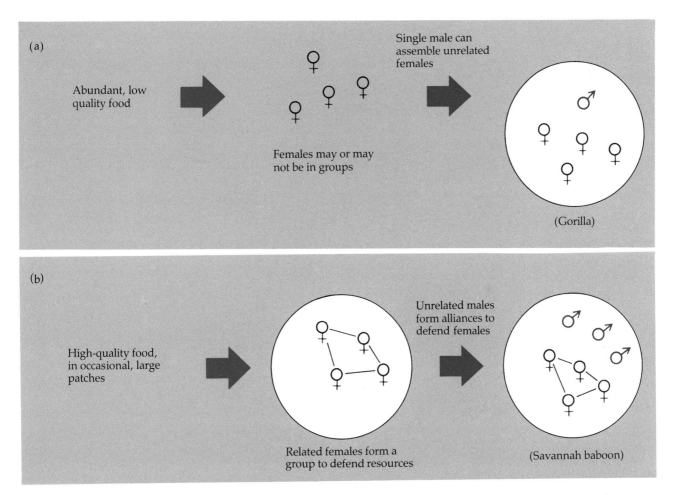

Distribution, with larger resource patches: In (a) when low-quality food is widely distributed, females may forage alone or in groups (in which the individuals are unrelated). A male may be able to assemble a harem, as does the gorilla. In (b) when high-quality food occurs in large but scarce patches, related females will form a group to defend them. Alliances among unrelated males may form, to defend the females from other males, as in savannah baboons.

The fact that there is such a rich array of social organizations among primates as a whole, and among the apes in particular, surely indicates that a rather complex set of processes underlies them. For each species there must be some kind of interaction between its basic phylogenetic heritage—its anatomy and physiology—and key factors in the environment. This means that different species will probably react differently to the same environmental factors, thus creating at least part of the observed diversity. What else plays a part?

'There is no consensus as to how primate social organization evolves,' Richard Wrangham of the University of Michigan observed recently, 'but a variety of reasons suggest that ecological pressures bear the principal responsibility for species differences in social behavior.' Indeed, for more than two decades ecological influences have been a popular source of explanations. The problem is,

explains Wrangham, that 'we do not know exactly what the relevant ecological pressures are, or which aspects of social life they most directly affect, or how'.

One of the most frequently advanced explanations of the benefits of group living has been defense against predation: even though it may be more conspicuous than a lone individual, a group can be more vigilant (more pairs of eyes and ears) and more challenging (more sets of teeth). Effective defense against predators has been observed in many group-living species of primate.

It is certainly true that terrestrial species, which are more at risk from predators than are arboreal animals, live in larger groups, commonly have more males in the group, and the males frequently are equipped with large, dangerous canine teeth. But for each of these three factors one can advance equally plausible explanations of their origin that have nothing to do with protection against predation. So, it is possible that terrestrial primates evolved these characteristics for these other reasons, and yet, once evolved, the properties are indeed effective against the threat of predation. Protection against predation may to some degree be a consequence, not the primary cause, of group living.

Food distribution has also been suggested as a trigger of social organization: groups might be more efficient than individuals at discovering discrete patches of food, for instance; or, where food patches are defensible by territorial species, then the patch size will influence the optimum group size. More recently, Wrangham has proposed a theory of social organization that includes food distribution as a key influence, but the focus of the model is different from earlier ideas.

The model examines the evolutionary context of male and female behavior, and proposes that 'it is selection pressures on female behavior which ultimately determine the effect of ecological variables on social systems'. In other words, whatever ecological setting a species might occupy, it is the behavior of females that is fundamental to the social system that evolves within it.

The reproductive success of female primates, as with all mammals, is determined by the number of offspring she can successfully raise: access to mature males is not usually a limiting factor, whereas access to food resources most certainly is. Male primates, in company with 97 per cent of all male mammals, bestow no parental care on their offspring: their reproductive success is therefore determined by successful access to mature females.

In the great majority of primate societies females remain in their natal group while males transfer. Any explanation of why primates should form social groups at all must also explain this asymmetry. Attempts to correlate different types of habitat with the tendency to form different types of social groups fail to do this. Wrangham's model does offer an explanation, as follows.

If food generally comes in patches that can support only one female and her offspring, then females will forage alone, as orangutan and chimpanzee females do much of the time. However, food that comes in larger, defensible patches can support several mature females and their offspring. Sharing a food resource also carries with it an element of competition within the group, which brings costs in terms of time and energy lost in aggressive encounters. Wrangham suggests that the costs of competition within a group is balanced against the benefits of cooperating with group members to outcompete other groups for access to food patches. Cooperation is most beneficial when it is among relatives: helping kin is like helping yourself, because they share your genes.

So, when a species exploits food resources that come in discrete, defensible patches, multifemale social groups will evolve, groups in which the females are closely related to each other. In anthropology such groups are known as matrilocal, but with nonhuman primates a better term is female-

Polygynous social system
Dimorphic canines

Monogamous social system
Monomorphic canines

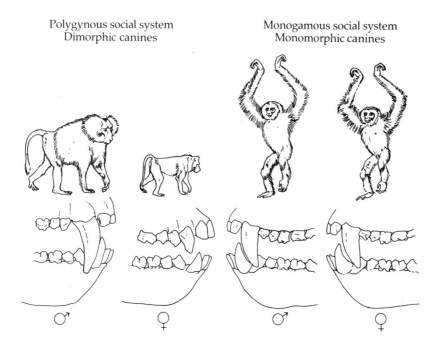

Sexual dimorphism, teeth and bodies: In polygynous social systems the males are typically larger than females, both in body size and in the canine teeth, as illustrated here by baboons. By contrast, in monogamous species, body size and canine size are usually very similar between the sexes, as illustrated here in gibbons. (Courtesy of John Fleagle/Academic Press.)

bonded. Where do the males fit in? If patches of food resources are relatively densely distributed, allowing a group of females to defend them all and be territorial, extra males are somewhat extraneous, and a unimale social system usually forms. If, however, territoriality is not possible and increased group size is not so much of a problem, several adult males can be accommodated. Indeed, extra males can be useful in the occasional competitive encounters with other groups. Here, some kind of multimale system would form.

In nonfemale-bonded systems, such as the chimpanzee and orangutan, where food does not come in defensible patches and females are mostly alone, the distribution of males depends on whether they can defend a community range alone or need the cooperation of other males. For orangutans, community denfence is possible by a single male, but for chimpanzee cooperation is necessary. Again, cooperation is most effective among relatives, and so chimpanzees have evolved a multimale social system in which females, not males, transfer to other groups on reaching maturity.

One possible consequence of primate sociality is body size, specifically the difference between males and females, known as sexual dimorphism in body size. Male primates often have to compete with other males for access to breeding females, and the bigger their body size, the more likely they are to succeed. By simple natural selection, in species in which such male–male competition occurs, male body size is likely to increase. Other factors that might be important in such encounters—canine teeth, for example—are also likely to become exaggerated in males.

It is certainly true that in monogamous species, in which competition between males is low or absent, males and females are typically the same size. It is also true that all species in which there is significant sexual dimorphism exhibit some degree of polygyny. Enlarged canines are also found in polygynous species. But the equation is not simple, because there is no direct correlation between the degree of polygyny and the degree of body size dimorphism: species in which males typically control harems of, say, 10 females do not necessary display greater body size dimorphism than species in which males control harems of two females.

Although the notion that body size dimorphism is the outcome of competition among males for access to females is popular among biologists, there are other possible explanations. The simplest is that males are large and aggressively equipped in order to provide effective protection against predators. But the old problem of circularity occurs again here. Another suggestion is that males and females are of different sizes as a way of exploiting different food resources, thus avoiding direct resource competition.

Robert Martin of the Anthropological Institute in Zurich adds an important note of caution here, saying that perhaps our explanations have been too male-oriented, seeking to explain why the male size has increased. Instead, he suggests, perhaps the size difference is the result of the females having become smaller. 'Smaller females may breed earlier,' he notes, 'selection for earlier breeding might explain the development of sexual dimorphism in at least some mammalian species.'

Even though many aspects of the interaction of species and their different environments remain to be fully worked out, one thing is clear: the complete social behavior of a species is the outcome of the mix of causes and consequences of individuals coming together to coexist in groups.

Key questions:

• Why has protection against predation been so popular an explanation as the cause of group formation in primates?
• Under what circumstances might male primates find themselves forced to contribute to the raising of offspring?
• If cooperating with one's kin indirectly benefits one's own genes, what advantages might there be in cooperating within non-kin in a social group?
• In considering fossil species—such as early hominids—what anatomical clues might be available about the social system of the living animals?

Key references:

P.H. Harvey and P.M. Bennett, 'Sexual dimorphism and reproductive strategies', in *Human sexual dimorphism*, edited by J. Ghesquiere, R.D. Martin, and F. Newcombe, Taylor and Francis, 1985.

L.A. Willner and R.D. Martin, 'Some basic principles of mammalian sexual dimorphism', *ibid.*

Richard Wrangham, 'Evolution of social structure', in *Primate societies*, edited by B. B. Smuts *et al.*, Chicago University Press, 1987.

12 / Primate models

Some scholars argue that evolution has carried humans so far away from the rest of the animal world that little can be learned about ourselves by looking at nonhuman primates. There is of course a gap between the behavioral repertoire of *Homo sapiens* and that of even our closest genetic relatives, the African apes. But, as Robert Foley of Cambridge University, England, has pointed out most clearly recently, that gap must have been filled to some extent by the behavioral repertoires of our ancestors, the various extinct species of hominid. It is therefore legitimate to use what information is available about the behavioral ecology of living primates in order to infer something about the behavior of our forebears.

For instance, it is important to know something of the social organization of the last common ancestor between hominids and apes and of the early hominids themselves: did they live in groups, and, if so, what was their size, and what was the ratio of mature males to mature females? It is also important to remember that when we use 'early hominids' as a collective term we are including at least half a dozen species. And if the experience of observing the behavior of modern ape species gives us a lesson, it is that we can expect different forms of social organization among different hominid species.

There are several ways in which modern primates can be used to model the lives of the extinct species. First, one can identify a living species that appears to match in some important ways some basic hominid characteristics and then seek lessons about behavior that might transfer to hominids. Second, one can be guided by phylogeny, and look only at the living African apes and humans, and seek commonalities of behavior that might therefore have been present in a common ape/hominid ancestor. Third, now that an understanding of behavioral ecology is beginning to develop (unit 11), it is becoming possible to infer from basic principles the social organization of hominid forebears.

The first of these three—the specific primate model—is the longest established approach. Several different species have at different times been offered as *the* most appropriate model, including the savannah baboon, the common chimpanzee, and, most recently, the pygmy chimpanzee. Although the baboon is a monkey, not an ape, and is therefore genetically only rather distantly related to hominids, baboons were attractive as models for early hominids because they share a similar habitat: bushland savannah. Living in troops with as many as 200 individuals, the savannah baboon offered a striking picture of the social life of our forebears. The troop is made up of mature females (often related to each other) and their offspring, and many mature males (unrelated to each other), which are larger than the females and equipped with impressively threatening canines. In other words, it is a multimale, female-bonded social organization.

So powerful was this image, and so well studied were these animals that Shirley Strum, a baboon-watcher herself, said recently: 'the "baboon model" had a disproportionate impact on our ideas about primates.'

The chimpanzee has also been proposed as a model for the last common ancestor and the early hominids, and for good reason: it is our closest genetic relative. One problem with this, as with all specific models, is the trap of the present: just as extinct species are likely to be unique anatomically and not some slight variant of a living species, so the behavior of extinct species is also likely to be unique. When, for instance, a chimpanzee model is proffered, 'an ape—human dichotomy is created', says Richard Potts, an anthropologist at the Smithsonian Institution, Washington DC. 'The problem with placing early hominids along a chimp—human continuum is that it precludes considering unique adaptations off that continuum.'

Potts points out that the dentition of the early hominid genus *Australopithecus*—large, thickly-capped cheek teeth set in robust jaws—is like neither chimpanzees nor humans: 'Thus, in this aspect of dental anatomy *Asutralopithecus* did not fall on the proposed continuum', notes Potts.

The most recent entry into the primate model stakes is the pygmy chimpanzee, placed there in 1978 by Adrienne Zihlman, John Cronin, Vincent Sarich, and Douglas Cramer. Randall Susman, of New York University at Stony Brook, is also a proponent of the model. The pygmy chimpanzee, which is now to be found only in a small area in Zaire, is strikingly similar in overall body

A catalogue of candidates: Several different species have at different times been nominated as instructive models for early hominid evolution: here we see the pygmy chimpanzee (*top left*), the common chimpanzee (*top right*), the savannah baboon, and the lion (a social carnivore).

Method	Phylogenetic comparison	Chimpanzee model	Chimpanzee model	Behavioral ecology	Behavioral ecology
Species	Common ancestor	A late prehominid	The earliest hominid	The earliest hominid	An early hominid
1 Closed social network	Yes	—	Yes	Yes	—
2 Party composition	?	Unstable	Unstable	Stable	Unstable
3 Females sometimes alone	?	Yes	Yes	No	Yes
4 Males sometimes alone	Yes	Yes	Yes	No	Yes
5 Female exogamy	Yes	—	Yes	No	—
6 Female alliances	No	—	No	Yes	—
7 Male endogamy	?	Often	Yes	No	Yes
8 Male alliances	?	—	Yes	—	—
9 Males have single mates	No	No	No	No	Yes
10 Length of sexual relationships	?	Short	Short	Short	Long
11 Hostile relations between groups	Yes	No	Yes	—	—
12 Males active in inter-group encounters	Yes	—	Yes	—	—
13 Stalking and attacking	Yes	—	—	—	—
14 Territorial defence	?	—	?	—	—

Ancestral social organization: Using different models, it is possible to determine those aspects of behavior that might have appeared in an ancestral species. In the phylogenetic comparison, each of the 14 questions asks if a particular aspect of behavior exists in *all* the modern African apes: if it does, then the likelihood is that this same behavior also appeared in the common ancestor with hominids. (Courtesy of Richard Wrangham.)

proportions to the earliest known hominid, *Australopithecus afarensis*. This species, *Pan paniscus*, may therefore be an even better model of early hominids than the common chimp, *Pan troglodytes*. Nevertheless, one always has to remember that even closely related species may exhibit distinctly different social structures when they occupy very different habitats.

In addition to primate models for hominid forebears, social carnivores have occasionally been said to be instructive. The case here is strict analogy with a supposed behavior: cooperative hunting. As cooperative hunting among hominids was probably a rather late evolutionary development (unit 20), this model has limited utility.

The second approach—phylogenetic comparisons—is rather more conservative, seeking only to identify basic shared behavioral characteristics among humans and African apes. The rationale, as explained recently by Richard Wrangham of the University of Michigan, is as follows: 'If [a behavior] occurs in all four species, it is likely (though not certain) to have occurred in the common ancestor because otherwise it must have evolved independently at least twice. If the four species differ with respect to a particular behavior, nothing can be said about the common ancestor.'

Wrangham examined 14 different behavioral traits—such as social group structure, male–female interactions, inter-group aggression, and so on—and found eight to be common to gorillas, the two chimp species and humans, while six were not shared. On this basis Wrangham infers that the common ancestor of hominids and African apes 'commonly had closed social networks, hostile, male-dominated intergroup relationships with stalk-and-attack interactions, female exogamy and no alliance bonds between females, and males having sexual relationships with more than one female'.

This ancestral suite, as Wrangham calls it, is merely a foundation upon which past social behavior can be constructed. But, for instance, it does seem to preclude the suggestion made in 1981 by Owen Lovejoy that the earliest known hominid, *Australopithecus afarensis*, was monogamous and nonhostile.

The third approach—reconstructing social organization from first principles of behavioral ecology—is the newest and most promising. Developed most thoroughly by Robert Foley, the technique seeks to establish the range of social structures that might have been available to hominid ancestors, and then determine how these might change in the face of changing environments.

The basis of the analysis is the recognition of phylogenetic constraint in ecological context: just as ancestral anatomy limits the paths of subsequent evolution, so too does ancestral social structure.

For instance, evolving from a multimale, nonfemale-bonded organization to a multimale, female-bonded structure is highly unlikely, because the intermediate steps would be inappropriate to prevailing conditions. In other words, only certain evolutionary pathways are available for ecologically-driven shifts in social organization. So, if you know where among the many possible social structures you start with an ancestral species, you can predict the nature of ecologically-driven social change, because you know what the available pathways are.

The phylogenetic context for hominids is, of course, the apes, and particularly the African apes. The social structures found among the apes varies greatly, ranging from solitary individuals among orangutans, through monogamous families in gibbons, single-male units with small numbers of unrelated females among gorillas, to complex fission–fusion communities of chimpanzees. 'Despite this variation, these social structures...constitute a limited set of social outcomes and are evolutionary closely related', observed Foley recently, in conjunction with Phyllis Lee, also of Cambridge University. In marked contrast with common monkey social organization, none of the ape social structures involves a core of related females.

Foley and Lee suggest that the most likely social structure in species ancestral to African apes and hominids is rather gorilla-like. Towards the end of the Miocene, about 10 million years ago, a steadily

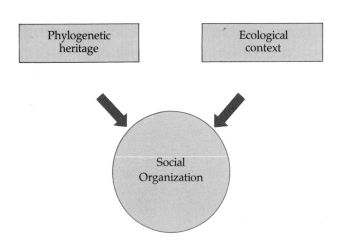

Contributions to social organization: A species' social structure will be determined by the outcome of interaction between its phylogenetic heritage—body size etc—and the environment in which it lives. Species with different phylogenetic constraints may therefore exhibit different social structures under the same environmental conditions.

cooling climate was reducing forest cover. A drier, more diverse habitat developed, especially in East Africa, which had the effect of making food resources more patchily distributed. Such an ecological shift would favor the evolution of a chimp-like social structure: communities of dispersed females and their offspring, with genetically related males defending the community against males from other groups.

'The emergence of the hominids can be seen as part of the African hominoid radiation, with this clade exhibiting increasingly strong male kin alliances under certain ecological conditions', say Foley and Lee. As the cooling continued the ecological shift would continue. 'Such environments appear to promote, among other things, larger group size among primates—partly as a response to the greater threat of predation, partly due to the effects of resources being more patchily distributed.' Given the evolutionary pathways available under the model, the larger social groups 'are more likely to be built upon the male kin alliances rather than related females'.

Given this background, Foley and Lee suggest that 'the most probable social organization for the early australopithecines...consists of mixed sex groups, with males linked by a network of kinship. Females, forced to forage over larger areas to find dispersed and seasonally limited food and to aggregate in the face of some predation, would be expected to form stable associations with either specific males within the alliance, or with the entire alliance of males.'

This social package, although apparently consistent with prevailing ecological conditions, is at variance with the fossil record of the earliest known hominid, at least in the way it is currently interpreted. This hominid, *Australopithecus afarensis*, is said to have a degree of body size dimorphism as extreme as that in orangutans. Dimorphism on this scale is usually taken to imply extreme competition between males for access to females. However, if, as Foley and Lee suggest, the males in these social groups are closely related to each other, then cooperation rather than fierce competition is to be expected. So, either the model is missing something, or the fossils currently interpreted as *Australopithecus afarensis* in fact represent two less dimorphic species.

Within the hominid species of 3 to 1 million years ago there developed a degree of morphological diversity, presumably reflecting different patterns of subsistence. At one extreme, the robust austra-lopithecines, a diet of coarse, low-quality plant foods was apparently exploited. Such foods tend to occur in large, widely dispersed patches. 'The expected effect on the ancestral hominid socioeco-logy...would have been to weaken male kin bonds within a less structured large or fluid group', note Foley and Lee. More competition between males would develop, presumably accompanied by dimorphism in body size.

At the other extreme, *Homo erectus*, the adaptation includes increased brain size and much reduced dental apparatus. Faced with the same problem of subsisting in tropical savannah environments— that is, maintain a constant food supply in the face of seasonality—this hominid adopted a strategy different from that of the robust australopithecines. Instead of exploiting low-quality food resources, they increased the consumption of meat, a patchily distributed but high-quality resource.

'The causes of meat-eating are ecological', note Foley and Lee. 'The consequences for the hominids would have been distributional and social.' One consequence would be greatly increased home and day ranges, which would make direct defense of females by males extremely difficult. 'The predicted response would be resource defense through ter-ritorial exclusion, and given the size of the area involved, this would involve alliances of males rather than individual defense. Such a pattern of behavior would enhance male kin associations as a means of coping with high levels of inter-group competition and inter-actions.'

In addition to changes wrought by this subsist-ence strategy, *Homo* would be facing another key change: the consequences of brain enlargement. Producing and rearing large-brained offspring is energetically expensive: at some point it would have become too expensive for the mother to pro-vide alone, at which point paternal involvement would become necessary. 'The effect...would be to increase the frequency, intensity, and stability of male–female associations', say Foley and Lee. Is this the beginning of the nuclear family, so much a part of Western society? No, because the nuclear family is in fact rather uncommon among human societies: an analysis of social structure variation among modern human societies as a whole shows that 74 per cent are polygynous.

The 20 per cent body size dimorphism in modern humans would indicate a degree of male–male competition in our recent past, not monogamy. And the fact that more resources are devoted to male fetuses than female fetuses, thus giving a higher birth weight, is also consistent with male–male competition. One further factor is the size of the male's testis, an indicator of subtle competition among males in multimale groups. For instance, chimpanzees live in promiscuous, multimale groups, and one way an individual male might outcompete his fellows to is produce more sperm in his ejacu-late. Gorillas and orangutans don't face this kind of competition, and they have small testes.

What of *Homo sapiens*? Human testes are small too, apparently ruling out competition in pro-miscuous, multimale groups. Monogamy appears to ruled out too, leaving a form of unimale polygyny. But, as Robert Martin and Robert May commented recently: 'these biological antecedents are today often overlain by extremely powerful socioeconomic determinants.'

Key questions:

• Why are extinct species likely to have dis-played unique behaviors?

• Why are 'first principle' approaches to infer-ring ancestral social behavior likely to be more difficult yet more informative than single species models?

• How important is phylogenetic constraint likely to be in the face of sharply changing en-vironmental conditions?

• What are the consequences of a male primate undertaking parental care through provisioning?

Key references:

R.A. Foley and P.C. Lee, 'Finite social space, evolutionary pathways and reconsidering hominid behavior', *Science* (in press).

R.D. Martin and R.M. May, 'Outward signs of breeding', *Nature*, vol 293, pp 7–9 (1981).

R. Potts, 'Reconstructions of early hominid sociecology: a criti-que of primate models', in *The evolution of human behavior: primate models*, edited by W.G. Kinzey, SUNY Press, 1987.

S.C. Strum and W. Mitchell, 'Baboon models and muddles', *ibid*.

R.W. Wrangham, 'The significance of African apes for recon-structing human social evolution', *ibid*.

A.L. Zihlman *et al.*, 'Pygmy chimpanzee as a possible proto-type for the common ancestor of humans, chimpanzees and gorilla', *Nature*, vol 275, pp 744–746 (1978).

13 / Hominid precursors

Humans belong to the infraorder Catarrhini, the group that includes Old World monkeys (superfamily Cercopithecoidea) and apes (superfamily Hominoidea). Humans are grouped with apes in the Hominoidea. The Catarrhini appears to have originated in Africa, with the evolutionary divergence between monkeys and apes probably occurring between 30 million and 20 million years ago. The human lineage arose some time between 10 million and 5 million years ago. In this unit we are concerned principally with the early apes, the group from which humans evolved.

Three key points stand out in any review of the evolution of the catarrhines.

First, the fossil record of the group for the most part does not overlap geographically with the areas where catarrhines are most abundant today. The early fossil record is concentrated in North Africa and Eurasia, with some in East Africa. Modern Old World monkeys and apes are most abundant in the forests of subsaharan Africa and southeast Asia. This pattern may reflect real changes in the history of the group. Or it may in part at least be the result of a biased fossil record: forest habitats are generally poor for fossil preservation.

Second, among living catarrhines, Old World monkeys are both more abundant and more diverse than apes: there are some 15 genera and 65 species of Old World monkey compared with five genera and a dozen species of hominoid. In the early history of the group the situation was precisely opposite, with apes being more abundant and more diverse than monkeys.

Third, the early apes were not merely quaintly primitive versions of the species we know today. They were combinations of various sorts of characters: some apelike, some monkeylike, and some that are unknown in modern large primates. One consequence of such anatomical novelties is that the early apes were probably behaviorally distinct too, in terms of locomotor patterns and dietary activities. Another consequence is that it is much more difficult to predict how various ancestral species might have looked and behaved, and this includes the ancestor of the human lineage.

If the current fossil record is a reasonable reflection of catarrhine history, then a number of rather general trends—in body size, brain size, and locomotor and dietary adaptation—can be discerned: such trends are common to most groups undergoing adaptive radiation.

First, there is an increase in body size among the group as a whole and within certain lineages in the group, particularly the apes. Second, relative brain size is generally bigger among the catarrhines compared with the prosimian primates. Third, the initial adaptive niche of quadrupedal, arboreal frugivory (fruit eating) broadens: modes of locomotion come to include suspensory climbing in trees (apes only), and terrestriality (apes and monkeys); leaf eating (folivory) becomes steadily more

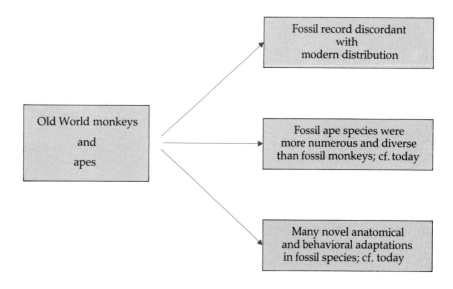

Features of the catarrhine fossil record: A major lesson to be inferred from these main features of the catarrhine fossil record is that the present is not always a direct key to the past.

important within the group as a whole (mainly monkeys).

Inevitably, the current fossil record is incomplete, as evidenced by the recent discovery of two new genera of Middle Miocene apes from East Africa. Our notion of the diversity of Miocene apes may therefore have to be further extended. The prospects of new fossil discoveries altering our picture of the major adaptive patterns of the group are, however, less likely.

The earliest known significant fossil record of Old World anthropoids occurs in the Fayum province of Egypt, arid badlands some 100 kilometers southwest of Cairo that 35 million years ago was a patchwork of moist riverine forest and woodland. During more than two decades of excavation at the site, Duke University paleontologist Elwyn Simons and his colleagues have recovered cranial and postcranial remains of about half a dozen primate genera, estimates of the body weights of which range between 0.3 kilogram for the smallest (*Qatrania*) to 6 kilograms for the largest (*Aegyptopithecus*). For the most part, the Fayum primates appear to have been monkeylike frugivores.

Although one of the species—*Aegyptopithecus zeuxis*—was once considered to be an early ape, all the Fayum species are now thought to antedate the evolutionary divergence of the anthropoid stock into Old World monkeys and apes. *Aegyptopithecus*, or something like it, such as its Fayum contemporary *Propliopithecus*, may represent that basic Old World anthropoid stock. However, some authorities consider it possible that *Aegyptopithecus* and its contemporaries might even represent the basic anthropoid condition prior to the split between Old World and New World anthropoids.

According to Simons, *Aegyptopithecus* was 'a generalized arboreal quadruped' with 'no evidence whatever...of either arm swinging or upright walking tendencies'. Behaviorally speaking, the closest living analogue to *Aegyptopithecus* is *Alouatta*, the howler monkey of the New World. 'This resemblance is due to similar adaptation and not genetic closeness', stresses Simons.

Just recently it became clear from the rapidly accumulating fossil collection from the Fayum that in at least three of the species—*Aegyptopithecus*, *Propliopithecus*, and *Apidium*—there is a significant difference in size between males and females. Drawing on patterns in modern species, this sexual dimorphism—specifically in the size of jaws and teeth—in which males were significantly larger than

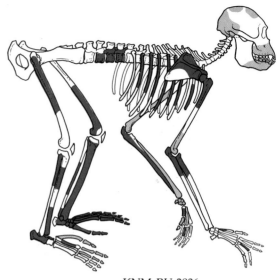

KNM-RU 2036

Proconsul africanus: This reconstruction is based on fossils found prior to 1959 (in blue) by Mary Leakey and in 1980, among the Nairobi Museum collections, by Alan Walker and Martin Pickford. The individual, a young female that lived about 18 million years ago, has characteristics of modern monkeys (in its long trunk and arm and hand bones) and of modern apes (in its shoulder, elbow, cranial and dental characteristics). (Courtesy of Alan Walker.)

females, can have implications for social structure (see units 10 and 11). For instance, the larger male size may be the result of competition between males for access to females within some kind of polygynous system.

After Fayum times, which samples a period somewhere between 37 million and 32 million years ago (the late Oligocene), the catarrhine fossil record is virtually nil until the Early Miocene (23 million to 17 million years ago). Here we must go to East Africa where there is evidence of the early ape group, most notably with fossils of *Proconsul*, a sexually dimorphic frugivore that lived about 18 million years ago, and the first evidence of Old World monkeys.

Proconsul is a very good example of the 'mosaic' nature of the early apes, as revealed by some unusually extensive skeletal remains recovered from western Kenya. 'In the forelimb skeleton, the shoulder and elbow region are remarkably apelike,' notes Alan Walker of Johns Hopkins University, 'but the arm and hand bones look more like those of some monkeys. In the hindlimb the reverse is true: the foot and lower leg bones are very apelike while the hip region looks less so.' The lumbar vertebrae are interesting too, looking more like those of gibbons than those of Old World monkeys or large apes.

This anatomical confection implies a mix of ape and monkey components of locomotion, a combination of quadrupedal running and leaping with vertical climbing and arm swinging. The confection also offers a salutary warning against sweeping interpretation based on limited fossil evidence: if the arm and hand bones were all that was known of *Proconsul*, then it would be thought to be a monkey; similarly, analysis based only on the foot and lower leg would imply ape, as would the shoulder and elbow.

The Early Miocene was a time of increasing diversity among protoapes, numbering at least half a dozen genera and more than twice as many species. These early apes were confined to Africa until about 18 million years ago, when the continent once again joined with Eurasia. Apes were among the many African species that took part in the subsequent faunal exchange between the two great landmasses, an event that was accompanied by further diversification as new geographic regions were occupied. It was close to this time that the lineage which eventually gave rise to the modern gibbon species diverged from the basic hominoid stock.

Rusinga Island in the Early Miocene: This community of apes, 18 million years ago, illustrates something of the species diversity that would later become the characteristic of monkeys. Upper left is *Proconsul africanus*; upper right, *Dendropithecus macinnesi*; center, *Limnopithecus legetet*; lower, *Proconsul nyanzae*. (Courtesy of John Fleagle/Academic Press.)

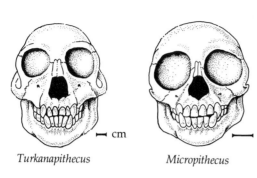
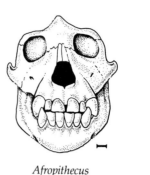
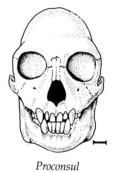

Turkanapithecus   Micropithecus   Afropithecus   Proconsul

Miocene apes: These faces of African apes, some 18 million years ago, illustrate the diversity of morphology among Miocene apes. (Courtesy of John Fleagle/Academic Press.)

Most Middle to Late Miocene (17 to 8 million years ago) hominoid fossils come mainly from Eurasia, although the East African record is rapidly improving. Debate ranges back and forth over how this diverse array of species might usefully be grouped. A decade ago it was common to see published phylogenies confidently linking modern hominoids with Middle Miocene fossil species, unbroken lines drawn through long periods of evolutionary time, even though intermediate fossil evidence was in virtually all cases totally absent. Researchers today are more cautious, acknowledging that only one living hominoid—the orangutan—can with any confidence be associated with a Middle Miocene ape—*Sivapithecus*.

Until very recently, some scholars divided middle Miocene hominoids into two groups: 'dryomorphs' and 'sivamorphs', based principally on characteristics of jaws and teeth. Dryomorphs, named after various species of *Dryopithecus* that lived in Europe, were said to be somewhat apelike (in the modern sense) in their jaws and teeth, and have sometimes been called dental apes. Sivamorphs, named after several Eurasian species of *Sivapithecus*, were grouped for their more robust jaws and teeth, which were capped with thick enamel. Estimates for the body weight of the several *Sivapithecus* species range between 20 and 90 kilograms, with sexual dimorphism being apparent in each of them.

Although not as well known postcranially as *Proconsul*, *Sivapithecus* can be described as most definitely hominoid in its locomotion, although no single label appears to be appropriate in terms of modern species. What appears to link *Sivapithecus* phylogenetically with the orangutan are various shared–derived homologies in the lower face and palate. However, as Pilbeam is careful to point out, '*Sivapithecus* is not an orangutan, nor is the orang-

utan a living fossil.' Apart from the shared facial characters, the orangutan appears to have evolved a long way from its Middle Miocene relative.

Other 'sivamorphs' include *Gigantopithecus*, an enormous Asian, ground-living ape that probably weighed more than 150 kilograms, and *Kenyapithecus*, which lived in Africa. Until recently, the group would also have contained the Eurasian species of *Ramapithecus*, but many investigators now consider this genus to be the same as *Sivapithecus*.

As it turns out, the neat 'sivamorph'/'dryomorph' division for the Middle Miocene is an artificial simplification. For instance, the thick enamel, robust dental structure that was thought to unite the 'sivamorphs' is much more complex than was imagined. As a result of this insight, the group is now recognized to have been much more heterogenous: 'They are not a natural group', as Pilbeam and his colleague Jay Kelly put it. The same qualification applies to the 'dryomorphs'.

The broad picture of hominoid evolution through the Miocene now begins to look like conglomerate series of repeated bursts of adaptive radiations, each covering some 2 to 3 million years. Making phylogenetic connections either within any one time period or vertically through time therefore becomes extremely challenging, if not impossible with the current patchy fossil record. Given the gaping fossil void that precedes the hominid fossil record and the even bigger hiatus preceding the modern African great apes, any guess as to the identity of the ancestor of the modern hominoids would be just that—a guess. For many people, that guess would lie in the direction of something that previously would have been called a 'sivamorph'.

For two decades since the early 1960s, *Ramapithecus* held a special place in paleoanthropology, being regarded by most researchers as the first hominid, giving human origins a date of at least 15 million years. Among the characters that were thought to link *Ramapithecus* with hominids were a robustly built jaw, with large cheek teeth capped with thick enamel, a set of characters that were considered to be hominid specializations. Several developments combined to depose *Ramapithecus* from its putative hominid status.

First, evidence from molecular biology implied that hominid origins were more like 5 million years ago, not 15 million (see unit 9). Second, thick enamel turned out to be a primitive character, not a hominid specialization. (In fact, enamel thickness is a much more complex character than was ever imagined—see unit 15.) Third, the undoubted hominid *Australopithecus afarensis* was in many ways more apelike than *Ramapithecus* was held to be, and yet, dated at 3 million years, lived long after *Ramapithecus*. And fourth—the clincher for most people—was the discovery in the early 1980s that *Sivapithecus* was phylogenetically linked to the orangutan.

The logic of this last point was as follows. *Ramapithecus* was known to be closely related to *Sivapithecus*. So, if *Sivapithecus* was in some way related to the orangutan, then so too is *Ramapithecus*. And yet, hominids are known to be closely related to the African apes, the chimpanzee and gorilla, and not the Asian ape, the orangutan. Ergo, *Ramapithecus* cannot be a hominid.

So, *Ramapithecus* was no longer regarded as a hominid, and very soon was not even known as *Ramapithecus*, being subsumed under *Sivapithecus*. The question then is, from what did hominids derive?

The hominoid fossil record between 7 million and about 4 million years ago is again virtually nil, in both Africa and Eurasia. This, coincidentally, was a time when, according to current evidence, hominoid (ape) diversity was shrinking and cercopithecoid (Old World monkey) diversity was increasing. But if fossil evidence were to be discovered of the immediate hominid precursor, what would it look like? This creature would of course also be the immediate precursor of chimpanzees and gorillas.

Dentally, guesses David Pilbeam, 'it would have had dimorphic canines, thick enameled and very dimorphic cheek teeth'. (Almost certainly it would have been principally a frugivore.) Sexual dimorphism would have extended to body size, with 'females weighing about 30 kg, males perhaps twice that', and with the polygynous social structure that

this implies. The body would be that of a general hominoid: 'Arms and legs would have been long, chest broad. The lumbar region is a problem: did it have six lumbar vertebrae like early hominids, five like gibbons, or three or four like large apes?'

We undoubtedly are talking about a principally arboreal creature, rather like a chimpanzee. But what of the mode of locomotion while it was on the ground? If recent molecular biological evidence were correct in uniquely linking humans and chimpanzees (see unit 9), then the protohominid/protochimpanzee might itself be a knuckle walker, as modern chimpanzees and gorillas are. Or, as Pilbeam puts it, 'It might have included knuckle walking in its positional repertoire'. In any case, the protohominid/protochimpanzee locomotor pattern must have been readily transformable into classical knuckle walking in one case and bipedalism in the other.

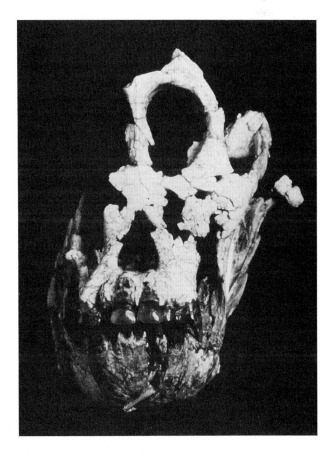

Sivapithecus: This 8 million-year-old specimen of *Sivapithecus indicus* comes from the Potwar plateau in Pakistan. The animal was about the size of a chimpanzee but had the facial morphology of an orangutan; it ate soft fruit (detected in the toothwear pattern) and was probably mainly arboreal. (Courtesy of David Pilbeam.)

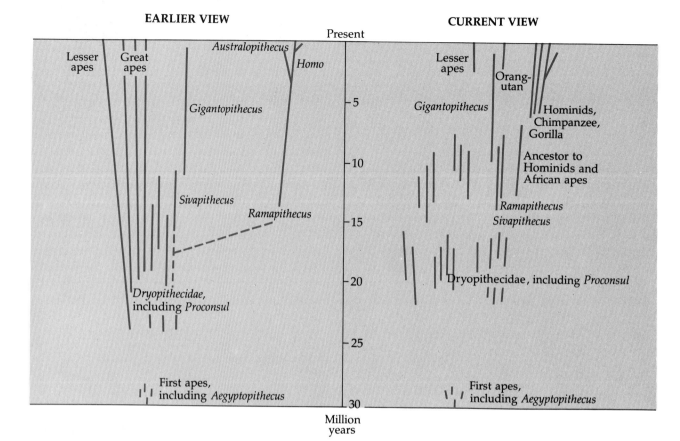

Changing ideas on hominoid ancestry: Earlier views on hominoid ancestry envisaged ladderlike progressions, with *Ramapithecus* branching off as the first hominid at least 15 million years ago. A large gap was placed between the hominids and the apes. Hominoid history, like the history of most animal groups, is now seen as a much more intricate bushlike structure. *Ramapithecus* is no longer thought of as the first hominid, and the gap between hominids and apes, particularly the African apes, has been closed.

Key questions:

● How influential has knowledge about extant cercopithecines been in shaping expectations about the history of the group?
● What is the most important lesson to be derived from an analysis of the anatomy and behavior of the Miocene hominoids?
● What evolutionary significance is implied by the concurrent decreased diversity of apes and increased diversity of the monkeys from the Middle Miocene onwards?
● With an improved fossil record in the appropriate time slot, how likely is it that the proto-hominid/protochimpanzee will be unequivocally identified?

Key references:

Peter Andrews, 'Hominoid evolution', *Nature*, vol 295, pp 185–186 (1982).
Peter Andrews and Lawrence Martin, 'Cladistic assessment of extant and fossil hominoids', *Journal of Human Evolution*, vol 16, pp 101–118 (1987).
John G. Fleagle, 'The fossil record of early catarrhine evolution', in *Major topics in ape and human evolution*, edited by Bernard Wood, Lawrence Martin, and Peter Andrews, Cambridge University Press, 1986, pp 130–149.
John G. Fleagle and Richard F. Kay, 'The paleobiology of the catarrhines', in *Ancestors: the hard evidence*, edited by Eric Delson, Alan R. Liss, 1985, pp 23–36.
Jay Kelly and David Pilbeam, 'The dryopithecines:taxonomy, comparative anatomy, and phylogeny of Miocene large hominoids', *Comparative primate biology, volume 1: systematics, evolution, and anatomy*, Alan Liss Inc., pp 361–411.
David Pilbeam, 'Hominoid evolution and hominoid origins', *American Anthropologist*, vol 88, pp 295–312 (1986).

14 / Origin of bipedalism

'Human walking is a risky business', British anthropologist John Napier once remarked. 'Without split-second timing man would fall flat on his face; in fact with each step he takes, he teeters on the edge of catastrophe.' Much of the anatomical adaptations—skeletal and muscular—to bipedalism have to do with maintaining balance so as to avoid this catastrophe.

Although *Homo sapiens* is not the only primate to walk on two feet—for instance, chimpanzees and gibbons often do when circumstances demand it—no other primate does so habitually or with what is known as a striding gait. The rarity of habitual bipedalism among primates—and among mammals as a whole—has given rise to the assumption that it is inefficient and therefore unlikely to evolve. As a result, anthropologists have often sought 'special'—that is, essentially human—explanations for the origin of bipedalism.

Closely associated with this is the insidiously seductive recognition that, once an ape is bipedal, its hands become 'freed'—freed to carry things, such as food, and to manipulate things, such as tools and weapons. So powerful is this notion that it has often been difficult to escape the assumption that bipedalism evolved *in order to* free the hands.

However, quite apart from its anthropocentric bias, in which, as we saw in unit 3, hominid origins are often interpreted as meaning human origins, the freed hands hypothesis might well be an example of the potential dangers of explaining the origin of a current structure or function in terms of

its current utility. The human brain, after all, is unlikely to have evolved in order that people might write symphonies or calculate baseball batting averages. So, although it is possible that upright walking evolved because of the advantages of carrying or making things with emancipated hands, other explanations must be explored too.

Human evolution is often cast in terms of four major novelties in relation to the basic hominoid adaptation: upright walking, reduction of anterior teeth and the enlargement of cheek teeth, elaboration of material culture, and a significant increase in brain size. However, as the current fossil and archeological records indicate, these novelties arose at separated intervals throughout hominid evolution: in other words, it shows a pattern of mosaic evolution.

Stone-tool making appears to have originated at about the same time as significant brain expansion, about 2.5 million years ago. The earliest hominid fossils known so far—from Ethiopia and Tanzania—are dated a million years earlier, and show significant adaptation to bipedalism in combination with a hominid dental pattern that has distinct apelike overtones. According to current estimates, these creatures—*Australopithecus afarensis*—lived approximately 3 million years after the origin of the hominid lineage, which is put at about 6.5 million years ago.

It is therefore possible that the first hominid species—assuming that something preceded *A. afarensis*—might have been apelike in all respects, apart from an adaptation to upright walking. In which case, bipedalism would be the primary hominid adaptation.

In this unit we will look at some of the mechanics of bipedalism, the ecological context in which it might have arisen, and the development of hypotheses that purport to account for its evolution.

The striding gait of human bipedalism involves the fluid flow of a series of actions—collectively,

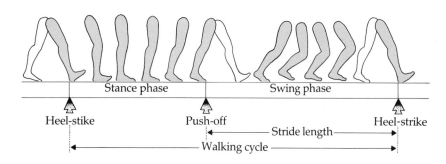

Phases of bipedalism: Upright walking in humans requires a fluid alternation between stance phase and swing phase activity for each leg. Key features are the push-off, using the great toe, at the beginning of the swing phase, and the heel-strike, at the beginning of the stance phase. (Courtesy of John Napier.)

Stance phase   Swing phase

Heel-stike   Push-off   Heel-strike

Stride length

Walking cycle

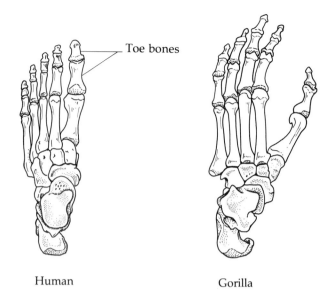

Toe bones

Human   Gorilla

Anatomy of the feet: The human foot is a platform, built for bipedalism, while the gorilla foot is more of a grasping organ. A key difference, therefore, is in the relationship of the great toe to the other toes of the foot: in humans the great toe is parallel with the other toes, in apes it is opposable. The foot bones of the earliest known hominid *Australopithecus afarensis*, have every appearance of being adapted to bipedalism; but the toe bones are also curved, which implies some arboreal activity.

the swing phase and the stance phase—one leg alternating with the other. The leg in the swing phase pushes off using the power of the great toe, swings under the body in a slightly flexed position, and finally becomes extended as the foot once more makes contact with the ground, first with the heel (the heel strike). Once the heel strike has occurred, the leg remains extended and provides support for the body—the stance phase—while the other leg goes through the swing phase, with the body continuing to move forward. And so on.

Two key features differentiate human and chimpanzee bipedalism.

First, chimpanzees are unable to extend their knee joints—to produce a straight leg—in the stance phase, which means that muscular power has to be exerted in order to support the body: try standing with your knees slightly bent and you'll get the idea. The human knee can be 'locked' in the extended position during the stance phase, thus minimizing the amount of muscular power needed for support of the body. The constantly flexed position of the chimpanzee leg also means there is no toe-off and heel-strike in the swing phase.

Second, during each swing phase the center of

gravity of the body has to be shifted towards the supporting leg (otherwise one would fall over sideways); and the tendency for the body to collapse towards the unsupported side is countered by contraction of the muscles (gluteal abductors) on the side of the hip that is in the stance phase. In humans, because of the inward sloping angle of the thigh to the knee (the valgus angle), the two feet at rest are normally placed very close to the midline of the body. Therefore, the body's center of gravity does not have to be shifted very far laterally back and forth during each phase of walking.

In chimpanzees, the thigh bone does not slope inwards to the knee as much as in humans, which means that the feet are normally placed well apart. Also the gluteal abductors are not highly developed. During bipedal walking the animal is therefore forced to shift its upper body substantially from side to side with each step so as to bring the center of gravity over the weight-bearing leg. This characteristic waddling gait is exacerbated by the fact that in chimpanzees the center of gravity is higher, relative to the hip joint, than it is in humans.

Chimpanzee anatomy is a compromise between an adaptation to tree climbing and terrestriality (mainly knuckle walking). Human anatomy is a fully terrestrial adaptation. As we shall see later, these differences have implications for energetic efficiency.

Not surprisingly, the suite of anatomical adaptations that underlie human bipedalism is extensive, and includes the following: a curved lower spine; a shorter, broader pelvis and an angled femur, which are served by reorganized musculature; lengthened lower limbs and enlarged joint surface areas; an extensible knee joint; and a platform foot in which the enlarged great toe is brought in line with the other toes. On the face of it, therefore, one might wonder how on earth a quadrupedal ape might possibly have undergone the required evolutionary transformation to produce a fully committed biped.

The nature of the evolutionary transformation that was required would of course have depended on the nature of the locomotor adaptation of the immediate hominid ancestor. The ancestor might have been a knuckle walker, like the chimpanzee, or instead much more arboreally adapted. In any case, the quadrupedal to bipedal transformation is not as dramatic a shift as it might at first appear.

'All primates, with one or two possible exceptions, can *sit* upright, many can *stand* upright without any support from their arms, and some can *walk* upright', notes Napier. 'In other words, we must view the human upright posture not so much as a unique hominid possession but as an expression of an ancient primate evolutionary trend....The dominant motif of that trend has been an erect body.' That trend went through vertical clinging and leaping (in prosimians), through quadrupedalism (in monkeys and apes), to brachiation (in apes). In other words, the transformation from ape to hominid was not between a true quadruped (such as a dog or a horse, for instance) and a true biped, a point that becomes important in calculating the evolutionary constraints that might have operated in the origin of hominids.

The earliest hominid fossils have been found in East Africa in what 3.5 million years ago would have been mosaic environments composed of interleafed forest, open woodland, and bush savannah, usually close to water. Unless this is pure taphonomic bias, the implication is that the earliest hominids might have evolved under similar eco-logical circumstances, which in general is a rather more open habitat than is typical for extant and extinct large hominoids. The popular notion of our forebears striding out of dense forest and onto open grassland savannah is, however, likely to be somewhat far fetched.

As we saw earlier (unit 3), Darwin essentially equated hominid origins with human origins, and proposed an evolutionary package that included upright walking, material culture, modified dentition, and expanded intelligence. For Darwin—and for many who followed—upright walking was required for the elaboration of culture: ...the hands and arms could hardly have become perfect enough to have manufactured weapons, or to have hurled stones and spears with true aim...so long as they were habitually used for locomotion and supporting the whole weight of the body.'

In the 1960s this incipient 'Man the hunter' scenario found an added advantage in bipedalism, with the realization that, although humans are slower

Adaptations to bipedal locomotion

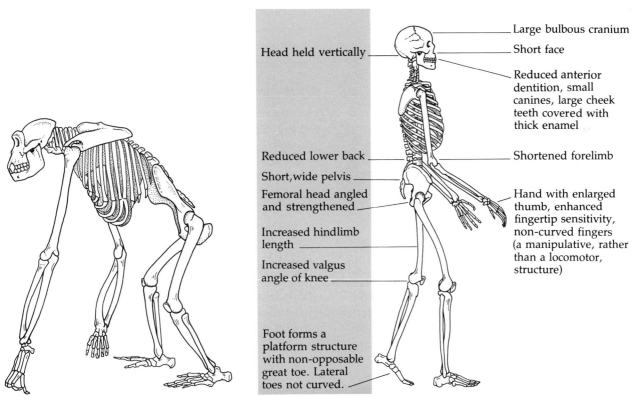

Head held vertically

Reduced lower back

Short, wide pelvis

Femoral head angled and strengthened

Increased hindlimb length

Increased valgus angle of knee

Foot forms a platform structure with non-opposable great toe. Lateral toes not curved.

Large bulbous cranium

Short face

Reduced anterior dentition, small canines, large cheek teeth covered with thick enamel

Shortened forelimb

Hand with enlarged thumb, enhanced fingertip sensitivity, non-curved fingers (a manipulative, rather than a locomotor, structure)

Skeletons compared: The ape, left, is adapted to a form of quadrupedalism known as knuckle walking, seen only in chimpanzees and gorillas. Human bipedalism, right, involves a number of anatomical differences from that seen in quadrupedalism, as indicated. Anthropologists are divided over whether or not the common ancestor of humans and African apes were knuckle walkers.

and less energy efficient than quadrupeds when running at top speed, at a slow pace it allows for great stamina such as might be effective in tracking and finally killing a prey animal. And just recently, with the 'Man the hunter' image being replaced by 'Man the scavenger' (unit 20), it has been suggested the endurance locomotion provided by bipedalism enabled the earliest hominids to follow in the wake of migrating herds, opportunistically scavenging the carcasses of the unfortunate young and the infirm old.

One problem with both these explanations is that, not only do stone tools that are required for getting access to meat from carcasses apparently postdate hominid origins by as much as 4 million years, but also there is no indication of regular meat eating in the dentition of the earliest known hominids. In fact, according to evidence from microwear patterns on the surface of teeth (unit 15), hominid diets remained predominantly vegetarian until about 1.5 million years ago, the origin of *Homo erectus*.

Other explanations offered for the origin of bipedalism have included the following: improved predator avoidance, by being able to see further across the 'open plain' on two legs than on four;

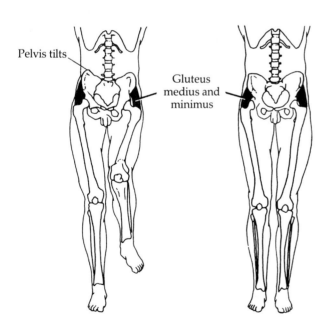

The pelvic tilt: Gluteus medius and minimus muscles link the femur (thigh bone) with the pelvis. They contract on the side in stance phase, thus preventing collapse to the side of the unsupported limb. Nevertheless, the pelvis tilts during walking. (Courtesy of David Pilbeam.)

display or warning; a shift in diet, such as seed eating; and carrying things. This latter has featured in two hypotheses in recent years, one being the 'Woman the gatherer' hypothesis, the other the 'Man the provisioner' model.

The 'Woman the gatherer' hypothesis was an openly ideological challenge to the male-oriented 'Man the hunter' model. Advanced initially in the early 1970s, the hypothesis shifted putative evolutionary novelties from males to females. 'Plants, not meat, were major food items, and plants, not meat, were the focus for technological innovation and new social behaviors', explains Adrienne Zihlman, an anthropologist at the University of California at Santa Cruz and the primary proponent of the 'Woman the gatherer' hypothesis.

The focal social ties envisaged here were those between females and their offspring, with the males being rather peripheral, a system that would have been a continuum with social structures in most other large primates (unit 11). The evolutionary novelty was that, living in a more open habitat than other large hominoids, hominid females had to travel substantial distances during foraging, sometimes using wooden tools to reach underground food items; they shared their food with infants; and often carried food and infants during foraging; hence the selective advantage of bipedalism.

The 'Woman the gatherer' hypothesis is more conservative than the 'Man the hunter' model, in that the first hominids are viewed as being basically apelike rather than already essentially human. Nevertheless, the hypothesis focuses on the need to carry things: specifically, food for sharing within infants.

The second recent hypothesis that focuses on the need to carry things is 'Man the provisioner', in which males gathered food and returned it to some kind of home base where it was shared with females and offspring, specifically 'his' female and offspring. Proposed in 1981 by Owen Lovejoy of Kent State University, the model envisages pair bonding and sexual fidelity between male/female couples, with the male providing an important part of the dietary resource.

The effect of such provisioning would be that the female would be able to reproduce at shorter intervals, thus giving them a selective advantage over other large hominoids, which, says Lovejoy, were reproducing at a dangerously slow rate. The system would work only if the males could be reasonably certain that the infants he was helping

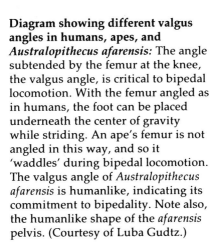
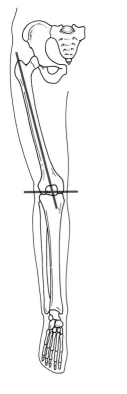

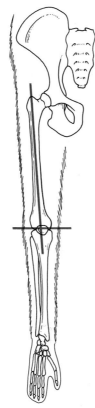

Diagram showing different valgus angles in humans, apes, and *Australopithecus afarensis:* The angle subtended by the femur at the knee, the valgus angle, is critical to bipedal locomotion. With the femur angled as in humans, the foot can be placed underneath the center of gravity while striding. An ape's femur is not angled in this way, and so it 'waddles' during bipedal locomotion. The valgus angle of *Australopithecus afarensis* is humanlike, indicating its commitment to bipedality. Note also, the humanlike shape of the *afarensis* pelvis. (Courtesy of Luba Gudtz.)

Human knee   *Afarensis* knee   Ape knee

to raise were his, hence the need for pair bonding and sexual fidelity.

Although it received widespread attention, Lovejoy's hypothesis has been keenly criticized. One line of criticism focused on the calculations that purported to show that large hominoids were at a reproductive disadvantage compared with humans. Another pointed out that if the first hominids were indeed monogamous, then the very large degree of sexual dimorphism in body size seen in these creatures would be very difficult to explain, given what is known about primate sexual systems and social structure (see units 10 and 11).

The most parsimonious—and therefore scientifically attractive—explanation of bipedalism proposed in recent years is that by Peter Rodman and Henry McHenry of the University of California at Davis. Very simply, they suggest that bipedalism might have evolved, not as part of a change in the *nature* of diet or social structure, but merely as a result of a change in the *distribution* of existing dietary resources. Specifically, in the more open habitats of the Late Miocene, hominoid dietary resources became more thinly dispersed in some areas, the continued exploitation of which demanded a more energy-efficient mode of travel: hence the

evolution of bipedalism. Here, the evolution of bipedalism concerns improved locomotor efficiency associated with foraging, and nothing else.

The proposal is based on a few simple points. First, although human bipedalism is less energy efficient than conventional quadrupedalism at high speeds, it is just as efficient—or more so—at walking speeds. Second, chimpanzees are about 50 per cent less energy efficient than conventional quadrupeds when walking on the ground, and this applies whether they are knuckle walking or moving bipedally. Therefore, noted Rodman and McHenry, 'there was no energetic rubicon separating hominoid quadrupedal adaptation from hominid bipedalism'.

For bipedalism to evolve among hominoids, all that was necessary was the presence of a selective advantage favoring improved energetic efficiency of locomotion. A more dispersed food resource could provide such a selection pressure. In other words, bipedalism was 'an ape's way of living where an ape could not live', noted Rodman and McHenry. 'It is not necessary to posit special reasons such as tools or carrying to explain the emergence of human bipedalism, although forelimbs free from locomotor function surely bestowed additional advantages to human walking.'

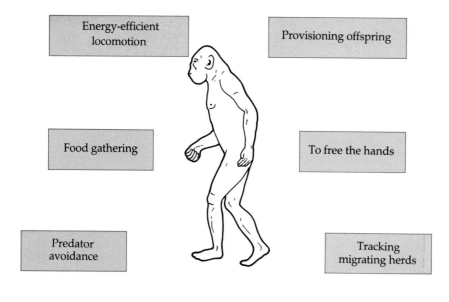

Hypothesized causes of bipedalism: Perhaps the defining characteristic of hominids, bipedalism has inevitably long been the focus of speculation as to its evolutionary cause. Some of the main ideas are shown here.

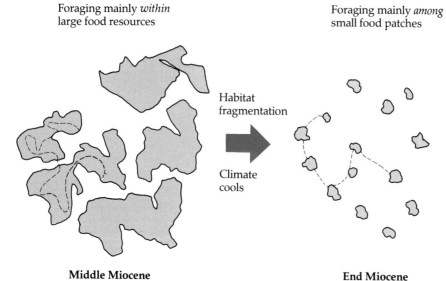

Foraging mainly *within* large food resources

Foraging mainly *among* small food patches

Habitat fragmentation

Climate cools

Middle Miocene

End Miocene

Habitat change: With the cooling and drying of the Late Miocene, once-continuous or near-continuous forest was fragmented in much of East Africa. Foraging for food resources would take place increasingly between patches rather than within them. This pattern of subsistence would therefore demand significantly further daily travel, thus producing selection pressure for more efficient locomotion. Under such conditions might bipedalism have evolved.

Key questions:

• What does the rarity of primate bipedalism imply, other than that it is 'difficult' to evolve?
• Given the energetic differences between hominoid quadrupedalism and human bipedalism, would the evolutionary transformation be *necessarily* fast or slow?
• How could one test the proposal that bipedalism evolved because of the advantages of carrying and/or manipulating things?
• Could a hominid that was completely apelike apart from being bipedal be classified as being a hominid?

Key references:

Clifford J. Jolly, 'The seed-eaters: a new model of hominid differentiation based on a baboon analogy', *Man*, vol 5, pp 1–26 (1970).
C. Owen Lovejoy, 'The origin of Man', *Science*, vol 211, pp 341–350 (1981). [See responses, vol 217, pp 295–306 (1982).]
Peter S. Rodman and Henry M. McHenry, 'Bioenergetics of hominid bipedalism', *American Journal of Physical Anthropology*, vol 52, pp 103–106 (1980).
Sherwood L. Washburn and C.S. Lancaster, 'The evolution of hunting', in *Man the hunter*, edited by Richard B. Lee and Irven DeVore, Aldine, 1968, pp 293–303.
Adrienne Zihlman, 'Women as shapers of the human adaptation', in *Woman the gatherer*, edited by Frances Dahlberg, Yale University Press, 1981, pp 75–119.

15 / Jaws and teeth

Jaws—particularly lower jaws—and teeth are by far the most common elements recovered from the fossil record. The reason is that, compared with much of the rest of the skeleton, jaws and teeth are very dense (and teeth very tough), which increases the likelihood that they will survive long enough to become fossilized.

Because jaws are usually an animal's principal food-processing machine, the nature of a species' dentition can yield important clues about its mode of subsistence and behavior. Overall, however, the dental apparatus is evolutionarily rather conservative, with dramatic changes being rare. For instance, human and ape dentition retains more or less the basic hominoid pattern established more than 20 million years ago. Moreover, different species under similar selection pressures as regards feeding habits may evolve superficially similar dental characteristics, as we shall see, for example, in the matter of enamel thickness. Similar sets of jaws and teeth may therefore belong to species with very different biological repertoires.

In this unit we shall examine four facets of hominoid dentition: first, the overall structure of jaws and teeth; second, the pattern of eruption; third, the characteristics of tooth enamel; and last, the indications of diet that are to be found in micro-wear patterns on tooth surfaces.

Perhaps the most obvious trend in the structure of the primate jaw (and face) throughout evolution is its shortening from front to back and deepening from top to bottom, going from the pointed snout of the tarsier to the flat face of *Homo sapiens*. Structurally, this change involved the progressive tucking of the jaws under brain case, giving a steady reduction of the angle of the lower jaw bone (mandible) until it reached the virtual 'L' shape seen in humans. Functionally, the change involved a shift from an 'insect trap' in prosimians to a 'grinding machine' in hominoids. Grinding efficiency increases as the distance between the pivot of the jaw and the tooth row decreases, hominids being closest to this position.

Old World anthropoids have a total of 32 teeth, arranged (in a half jaw) as two incisors, one canine, two premolars, and three molars. (New World monkeys have an extra premolar, giving a total of 36 teeth.) Overall, the modern ape jaw is rather rectangular in shape, while in humans it is more like an arc. But one of the most striking differences is that in apes the conical and somewhat blade-shaped canine teeth are very large and project way beyond the level of the tooth row; and male canines are substantially bigger than females', an aspect of sexual dimorphism with significant behavioral consequences (see unit 11).

When an ape closes its jaws the large canines are accommodated in gaps (diastemata) in the tooth rows: between the incisor and canine in the upper jaw and the canine and first premolar in the lower jaw. As a result of the canines' large size, an ape's jaw is effectively 'locked' when closed, side-to-side movement being prevented. By contrast, human canines—in both males and females—are small and barely project beyond the level of the tooth

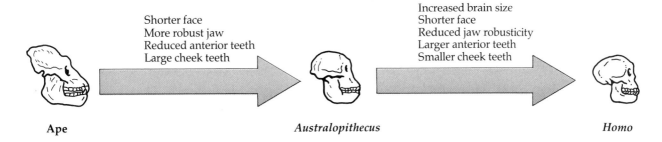

Ape   Shorter face / More robust jaw / Reduced anterior teeth / Large cheek teeth   *Australopithecus*   Increased brain size / Shorter face / Reduced jaw robusticity / Larger anterior teeth / Smaller cheek teeth   *Homo*

Evolutionary trends in dentition: The transition from ape to *Australopithecus* and from *Australopithecus* to *Homo* involved some changes that were continuous, some that were not. For instance, the face became shorter and shorter throughout hominid evolution, while robusticity of the jaw first increased and then decreased. The combined increase in cheek tooth size and decrease in anterior tooth size that occurred between ape and *Australopithecus* was also reversed with the advent of *Homo*.

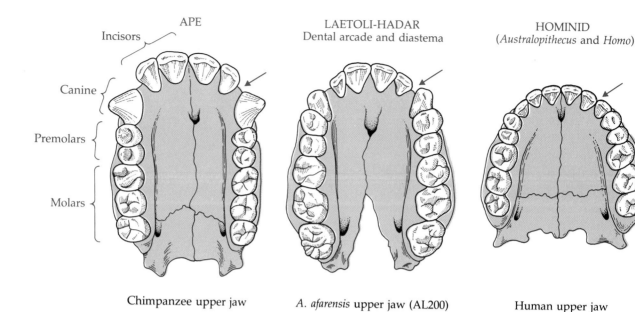

APE

Incisors

Canine

Premolars

Molars

Chimpanzee upper jaw

LAETOLI-HADAR
Dental arcade and diastema

A. afarensis upper jaw (AL200)

HOMINID
(*Australopithecus* and *Homo*)

Human upper jaw

Comparison of dentition in the palates of an ape, human, and *Australopithecus afarensis*: *A. afarensis* dentition is a mixture of hominid and ape characteristics. The incisors are relatively large, like an ape's; and in 45 per cent of specimens there is a gap—a diastema—between the canine and incisor. Such a gap is uncommon in later hominids. The canines are not as large as in apes, but neither are they as small as in some of the later hominids. The premolars are more primitive than those of later hominids, but the molars are very hominidlike in being large and usually showing a lot of wear to the cusps that generally makes the teeth rather flat. The cheek teeth in this specimen are in a straight line, like those of an ape, except for the last molar, which turns in a little to give a curve to the row. Hominid tooth rows from later taxa are generally arched as shown. (Courtesy of Luba Gudtz.)

row. As a result, the tooth rows have no diastemata, and a side-to-side 'milling' motion is possible, which further increases grinding efficiency.

In the earliest known hominids—*Australopithecus afarensis* from about 3.5 million years ago (see unit 16)—the canines are still relatively large and ape-like, with a significant degree of sexual dimorphism. Within a million years, however, the canines in several hominid species have become smaller and flattened, looking very like incisors (see unit 17).

The upper incisors of apes are large and spatula-like, which is a frugivore adaptation. Human upper incisors are smaller, more vertical, and, with the small, flat canines, form a slicing row with the lower teeth.

The single-cusped first premolar of apes is highly characteristic, particularly the lower premolar against which the huge upper canine slides. Ape molar teeth are larger than the premolars, and have high, conical cusps. In humans the two premolars are the same shape as each other, and have become somewhat 'molarized'. The molars themselves are large and relatively flat, with low, rounded cusps, characteristics that are extremely exaggerated in some of the earlier hominids (see unit 17).

The hominid dental package as a whole can therefore be seen as an extension of a trend towards a more effective grinding adaptation.

The pattern of eruption of permanent teeth in modern apes and humans is distinctive, as is the overall timing. Just recently anthropologists have been debating this aspect of hominoid dentition, specifically asking how early hominids fit in: were they more like humans or more like apes? Although the issue remains to be fully resolved, there are indications that until rather late in hominid history, dental development was in many ways rather ape-like, particularly in its overall timing.

The ape tooth eruption pattern is: M1 I1 I2 M2 P3 P4 C M3; and in humans: M1 I1 I2 P3 C P4 M2 M3. The principal difference, therefore, is that in apes the canine erupts after the second molar while in humans it precedes it. However, associated with the prolonged period of infancy in humans is an elongation of the time over which the teeth erupt. In apes the three molars appear at about 3.3, 6.6, and 10.5 years, whereas in humans the ages are 6, 12, and 18 years.

So, if you are faced with a human jaw in which the first molar has recently erupted, you can say that the individual was about 6 years old. An ape's jaw with the first molar just erupted would be from an individual a little more than 3 years old. The question is, if you have an early hominid jaw in this state, how old was it: 3 years or 6 years? As it happened, the first australopithecine to be discovered—the Taung child, *Australopithecus africanus*—was at just this state of development.

University of Michigan anthropologist Holly Smith recently analyzed tooth eruption patterns in a series of fossil hominids and concluded that most of the early species were distinctly apelike. For *Homo erectus*, which lived from 1.5 million until about 400,000 years ago, the results were somewhat equivocal, but with strong apish overtones. The human pattern was apparent in a Neanderthal child who died about 60,000 years ago.

Smith's conclusion has been challenged by University of Pensylvania anthropologist Alan Mann, who a decade earlier had proposed that all hominids followed the human pattern of development. Nevertheless, Smith's position received support in late 1987 when Glenn Conroy and Michael Vannier of Washington University School of Medicine published results of computerized tomography (CT) analysis of the Taung child's skull. Conroy and Vannier were able to 'see' the unerupted teeth within the jaw bone, and concluded that the teeth would have emerged in an apelike pattern.

The debate has been extended further by two researchers at University College London, who claim to be able to determine the exact age of a tooth by counting the number of lines—striae of Retzius—within the enamel. Although it is by no means universally accepted, the two researchers, Timothy Bromage and Christopher Dean, believe that the lines represent weekly increments—thus giving them an anthropological equivalent of tree rings, which measure yearly increments.

Bromage and Dean applied their technique to a series of australopithecine and early *Homo* fossils, and obtained ages that were between half and two-thirds of what would be inferred if a human standard of dental development had been applied.

If Smith and Bromage and Dean are correct, it seems that until relatively recently in evolutionary history, hominids followed a distinctly apelike pattern of dental development. This is important for its implication about the period of infant care. Once infant care becomes prolonged, which becomes necessary when postnatal brain growth is

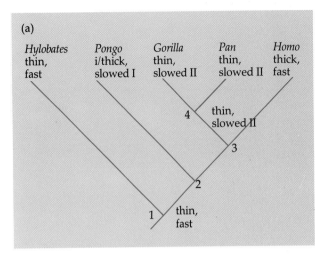

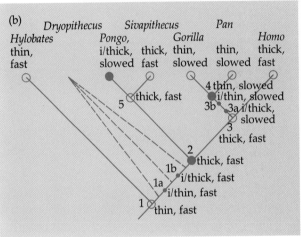

Patterns of tooth enamel formation: Tooth enamel was once thought to be of two patterns only: thin, as in apes; and thick, as in hominids. But the situation is more complicated. For instance, a thin enamel layer may be formed quickly (*Hylobates*, in (a)), or slowly (*Gorilla* and *Pan* in (a)). Equally, thick enamel may be formed quickly (as in *Homo*) or slowly (as in *Pongo*). By analyzing the details of different processes of enamel formation, and the possible evolutionary transitions between them, it is possible to devise an evolutionary tree, showing intermediate forms and places where fossil species might fit in (b). Closed circles at nodes indicate derived conditions with respect to previous ancestral node; open circles indicate nodes at which the ancestral condition is retained primitively. (Courtesy of Lawrence Martin/*Nature*.)

significant (see unit 27), then social life becomes greatly intensified. The dental evidence indicates that this prolongation may have begun with *Homo erectus*, which is in accord with data on increased brain size.

The relative thickness of tooth enamel has played an important role in anthropology, but it has been

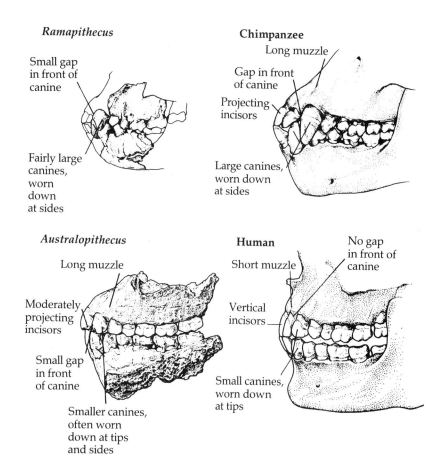

Ramapithecus

Small gap in front of canine

Fairly large canines, worn down at sides

Chimpanzee

Long muzzle

Gap in front of canine

Projecting incisors

Large canines, worn down at sides

Australopithecus

Long muzzle

Moderately projecting incisors

Small gap in front of canine

Smaller canines, often worn down at tips and sides

Human

Short muzzle

No gap in front of canine

Vertical incisors

Small canines, worn down at tips

Tooth characteristics: The diagram shows some of the major characteristics in (a) a ramapithecine, (b) a chimpanzee, (c) *Australopithecus afarensis*, and (d) *Homo sapiens*. (From *Our Fossils Ourselves*, courtesy of the British Museum (Natural History).)

recently discovered that this dental characteristic is much more complex than was once thought. Modern humans have a thick enamel coat on their teeth, as do all known fossil hominids, whereas the African apes have thin enamel. Thick enamel was therefore assumed to be an 'advanced' (or derived) character, and this was important in the argument that *Ramapithecus* was the first hominid (see unit 13). Similarly, thin enamel was assumed to be primitive. Thin enamel was seen as an adaptation to fruit eating, thick enamel a response to processing tougher plant foods.

The thin/thick enamel dichotomy is, however, misleading, as State University of New York anthropologist Lawrence Martin discovered. The thickness of an enamel layer depends on two factors: the speed at which it is deposited, and the length of time over which deposition occurs.

Enamel may be deposited in a fast mode, which produces a characteristic pattern 3 structure in the enamel prisms, and is primitive for hominoids. The slow mode of deposition produces pattern 1, which is derived among hominoids. In gibbons, for instance, a relatively brief burst of pattern 3—fast—deposition produces a thin enamel cap. A longer duration of the same mode of deposition in humans leaves a thick enamel layer. Now, chimpanzees and gorillas, like gibbons, have thin enamel, but deposition proceeds in two stages. Sixty per cent of deposition is in the fast mode, which is followed after a dramatic switch by slow deposition. Martin terms this pattern thin, slowed growth, which is developmentally and phylogenetically distinct from the thin, fast pattern in gibbons.

Orangutans, which have intermediate thick enamel, also go through a two-stage deposition, but again it is not homologous with that in African apes. After the initial fast phase (80 per cent of the total), deposition slows to 2.5 microns per day for about 200 microns, and then slows again to the African apes' lower rate for the final 50 microns.

A phylogenetic picture emerges, into which the fossil ape *Sivapithecus* fits neatly. This creature, which lived in Eurasia 15 to 8 million years ago and includes the forms known as *Ramapithecus*, turns out to have thick, fast-forming enamel, like humans. On the basis of facial anatomy, *Sivapithecus* is considered to be related to the orangutan.

Overall, then, the primitive dental pattern in hominoids is thin, fast-forming enamel, a condition represented today by gibbons. A lengthening of deposition time produced the derived state of thick, fast-forming enamel, seen in *Sivapithecus* and humans. Secondary slowing occurred in the orangutan, while the switch to the more complex thin, slowed pattern of the African apes appears to be most derived of all. Martin considers the sharing of the thin, slowed pattern between chimpanzees and gorillas to indicate that they derived from a common ancestor, and are not independently ancestral as indicated by certain molecular evidence (see unit 9).

The surface of tooth enamel is an animal's primary contact with its food, and to some extent at least a signature of that contact is left behind. Using a scanning electron microscope, Alan Walker of Johns Hopkins Medical School has produced images of a range of characteristic toothwear patterns: for grazers, browsers, frugivores, bone-crunching carnivores etc. In a series of comparisons, all early hominids appear to fit into the frugivore category, along with modern chimpanzees and orangutans. This pattern is a rather smooth enamel surface into which are etched a few pits and scratches.

A major shift occurs, however, with *Homo erectus*, whose enamel is heavily pitted and scratched. Such a pattern is most like a cross between a hyena (a bone-crunching carnivore) and a pig (a rooting omnivore). Although it is not yet possible to interpret precisely the implications for *Homo erectus* diet, it is significant that toothwear patterns indicate something of an abrupt change in hominid activities at this point in history. As will be seen in unit 21, *Homo erectus* chalked up a number of 'firsts'

in hominid history: significant brain expansion, reduction in body size dimorphism, systematic tool making, use of fire, migration out of Africa.

We have seen, therefore, that in spite of their limitations, teeth have the ability to yield information about hominid history that goes far beyond what simply went down our ancestors' throats.

Key questions:

- How reliable are teeth as indicators of a species' diet?
- What other information would one need in order to assess the significance of the reduction of overall size and loss of sexual dimorphism in hominid canines?
- How would you recognize the jaws and teeth of the first hominids?
- In evolutionary terms, how readily are enamel deposition patterns altered?

Key references:

Timothy G. Bromage and M. Christopher Dean, 'Re-evaluation of the age at death of immature fossil hominids', *Nature*, vol 317, pp 525–527 (1985).

Glenn C. Conroy and Michael W. Vannier, 'Dental development of the Taung skull from computerized tomography', *Nature*, vol 329, pp 625–627 (1987).

Alan E. Mann *et al.*, 'Maturational patterns in early hominids', *Nature*, vol 328, pp 673–675 (1987).

Lawrence Martin, 'Significance of enamel thickness in hominoid evolution', *Nature*, vol 314, pp 260–263 (1985).

B. Holly Smith, 'Dental development in *Australopithecus* and early *Homo*', *Nature*, vol 323, pp 327–330 (1986)

16 / The first hominids

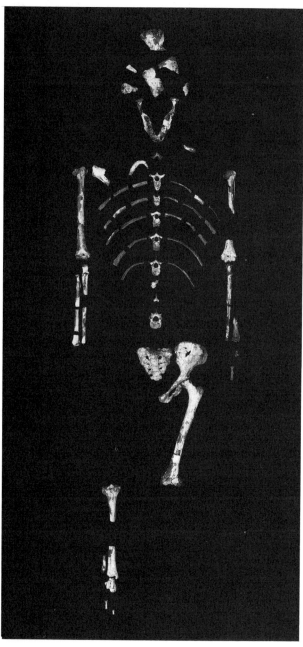

Skeleton of 'Lucy': This 40 per cent complete skeleton, one of the smallest specimens of *Australopithecus afarensis*, shows the combination of ape and human characteristics. Obviously adapted for considerable bipedalism, Lucy nevertheless had somewhat apelike limb proportions (short legs), and an apelike cranium and dentition. (Courtesy of the Cleveland Museum of Natural History.)

The earliest known hominid species is *Australopithecus afarensis*, the principal collections of which come from sites in Ethiopia (the Hadar, dated between 2.9 and 3.5 million years) and Tanzania (Laetoli, dated at about 3.6 million years). Older, more fragmentary remains that are clearly similar to if not identical with *A. afarensis* have been discovered from various parts of East Africa. The oldest, from Lothagam in Kenya, is dated at about 5.6 million years.

The Hadar and Laetoli specimens were recovered during the 1970s, and their interpretation quickly had a major impact on paleoanthropology. The site of Laetoli was especially significant because it also contains 3.6 million-year-old footprint trails of three hominid individuals, presumably *A. afarensis*. The footprints confirmed what was expected of early hominids and what was apparent in their postcranial skeleton: they were essentially bipedal, not quadrupedal animals. The site of Hadar is, however, the more important, not only because it yielded many more specimens (the remains of 65 individuals as against 14 from Laetoli), but also because one set of fossils constitutes about 40 per cent of the skeleton of one individual, the famous Lucy.

The initial interpretations of the Hadar and Laetoli fossils was somewhat diverse. For instance, in their 1976 paper on the Laetoli hominids, Mary Leakey, Tim White and their colleagues noted a 'phylogenetic affinity to the genus *Homo*'. In the same year, Donald Johanson and Maurice Taieb, joint leaders of the American/French expedition to the Hadar, reported that there might be three hominid species in the collection, two australopithecines and one *Homo*.

After a collaborative analysis, however, Johanson and White concluded that collectively the Hadar and Laetoli fossils represented just one hominid species, which they named *Australopithecus afarensis* in 1978. This was the first major hominid species to be named since 1964, when *Homo habilis* was announced. As with *H. habilis*, the naming of *A. afarensis* generated considerable controversy among paleoanthropologists.

Johanson and White described *A. afarensis* as being much more primitive than other known hominids, looking strongly apelike above the neck and strongly humanlike below the neck; as having extreme sexual dimorphism in body size (males bigger than females); and as being ancestral to all later hominids.

This position was challenged by some authorities,

most notably Richard Leakey and Alan Walker and by Yves Coppens and his colleagues in Paris. These critics argued that both the extreme size variation among the specimens, and many differences in detailed anatomy, implied the existence of at least two and perhaps more species in the collections. By now most scholars accept the interpretation of just one species—*A. afarensis*—in the collections.

Superficially, *A. afarensis* does indeed appear to be essentially apelike above the neck and essentially humanlike below the neck. This is a good example of mosaic evolution, in which different parts of the body change at different rates and at different times. In fact, mosaicism is even more pervasive and detailed in this species, because, throughout the postcranial skeleton, anatomy associated with bipedal locomotion is developed to different degrees in different places. One of the continuing debates over this species concerns the interpretation of the various primitive aspects of the postcranial anatomy: do they imply that, like most hominoids, *A. afarensis* still spent a significant amount of time in the trees? Or, were these primitive aspects of the anatomy simply genetic holdovers from an earlier adaptation, having no particular behavioral significance in *A. afarensis*? And while individuals were on the ground, was their bipedalism significantly different from or essentially the same as that in modern humans?

In examining the biology of *A. afarensis* in more detail, we will look first at the cranial and dental anatomy and then return to functional interpretations of the postcranial skeleton.

First, the cranial capacity of *A. afarensis* ranges between 380 and 450 cm^3, which is not much bigger then the 300 to 400 cm^3 span for chimpanzees. The cranium itself is long, low and distinctly like that of an ape, having a pronounced ridge (the nuchal crest) at the back to which were attached powerful neck muscles that balanced the head. Again as in apes, the upper part of the *A. afarensis* face is small, the lower part large and protruding. The projecting (prognathous) lower face is one reason why powerful neck muscles are required in order to balance the head atop the vertebral column: in physical terms, it is a matter of moments.

There are many details of the underside of the *A. afarensis* cranium (the basicranium) that signify its hominid status, including the central positioning of the foramen magnum, through which the spinal cord passes. But the hominid status of *A. afarensis* is more clearly seen in the jaws and teeth.

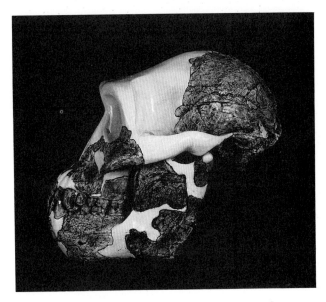

Reconstruction of afarensis cranium: The apelike features of *Australopithecus afarensis* are particularly evident in the cranium. The increased robusticity in the jaws, the slightly enlarged cheek teeth and reduced canines are major clues to its hominid status. (Courtesy of the Cleveland Museum of Natural History.)

A comparison of a modern ape's dentition (that of a chimpanzee, for example) with that of modern humans reveals some striking differences (unit 15). In most respects *A. afarensis* is somewhat intermediate between these two patterns. Although reduced, the canines are still large for the typical hominid, and there is significant sexual dimorphism; a diastema is required to accommodate each canine in the opposite jaw. In many individuals the first premolar is distinctly apelike in having a single cusp, but the development of a second cusp can sometimes be discerned. Although the molars are characteristically hominid in overall pattern, they are not the grinding millstones that are apparent later in the hominid lineage.

Clearly, an adaptive shift had occurred with *A. afarensis*, one that looks dramatic by comparison with modern apes. However, the proper comparison is with some of the Miocene apes (unit 13) from which the hominid clade might have derived. From this viewpoint, *A. afarensis* really does look apelike, both anatomically and dietarily too. However, behaviorally it would have been different, because *A. afarensis* had the ability habitually to walk on two legs. For an insight into this novel adaptation we move to the postcranial skeleton.

As we saw in unit 14, the bipedal adaptation imprints itself in many different ways on the postcranial skeleton. The question here is, how well does *A. afarensis* measure up as a biped? Functional analysis of various parts of the postcranial skeleton has been carried out by a large number of researchers, in the United States, England, and France.

Owen Lovejoy of Kent State University collaborated with Johanson and his colleagues, and concentrated on the pelvis and lower limbs. There is no doubt that the pelvis is more like that of a hominid than that of an ape, being squatter and broader, but there are significant differences too, such as the angle of the iliac blades (hip bones). These differences were not functionally significant in terms of achieving the balance required for

bipedal locomotion, concluded Lovejoy. And combined with the architecture of the femoral neck and the pronounced valgus angle of the knee, it would allow a full, striding gait, essentially like modern humans in overall pattern if not in every detail. In other words, *A. afarensis* was said to be a fully committed terrestrial biped, any apelike anatomy being genetic baggage and not functionally significant.

Meanwhile, other researchers began to see indications of arboreal adaption in the *A. afarensis* anatomy. French researchers Christine Tardieu and Brigitte Senut studied the lower limb and upper limb respectively, and inferred a degree of mobility

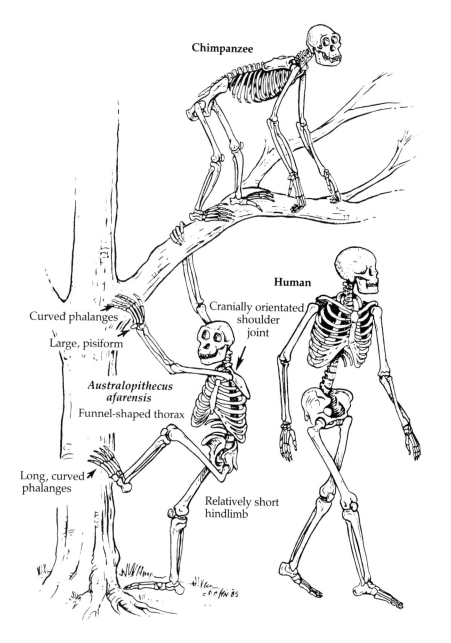

Skeletons compared: This diagram illustrates the skeletal adaptations to arboreality in *Australopithecus afarensis*. (Courtesy of John Fleagle/Academic Press.)

that would be consistent with arboreality. Russell Tuttle of the University of Chicago pointed out that the bones of the hands and feet were curved like an ape's, which could be taken as indicating climbing activity. William Jungers of the State University of New York (SUNY) at Stony Brook reported that although *A. afarensis* arms are hominid-like in length, the legs remain short, like an ape's, which favors a climbing adaptation. Henry McHenry of the University of California at Davis examined certain *A. afarensis* wrist bones, and concluded that the joint would have been much more mobile than in modern humans, again consistent with an arboreality.

Following a more wide-ranging survey, Jungers, Jack Stern and Randall Susman, also of SUNY, argued that the full suite of postcranial anatomical adaptations indicated that although *A. afarensis* was bipedal while on the ground, it spent a significant amount of time climbing trees, for sleeping, escaping predators, and foraging. Moreover, they concluded that while the animal was moving on the ground it was unable to achieve a full striding gait, as Lovejoy had argued, but instead adopted a bent-hip, bent-knee posture. Such a posture would clearly have important biomechanical and energetic implications for *A. afarensis*.

The differences of opinion in the *A. afarensis* locomotor debate stem partly from a lack of agreement over what exactly the anatomy is in certain instances, and differences in functional interpretation of other aspects of the anatomy. The opposing views were aired on an equal footing at a scientific symposium organized by the Institute of Human Origins in Berkeley in 1983. Since that time most publications have favored the partially arboreal, bent-hip, bent-knee bipedal locomotor posture.

The key anatomical features cited in support of a partially arboreal adaptation are several, and include the following: curved hand and foot bones, great mobility in the wrist and ankle, a shoulder joint (the glenoid fossa) that is oriented towards the head more than in humans, and short hind limbs. Opponents of arboreal adaptation dispute the degree of mobility in the *A. afarensis* ankle, and cite the loss of the opposable great toe, which has become aligned with the other toes, a clear adaptation to bipedality (but see below).

Anatomical features that might imply a less than human style of bipedality are to be found in several parts of the body. For instance, although the forelimbs have assumed hominid proportions, thus improving weight distribution and balance required for bipedalism, the legs are short, like an ape's. Short legs mean short stride length. Not only that, but the foot is long relative to the leg, meaning that clearance could be achieved only by increasing knee flexion during walking (like trying to walk in oversized shoes).

The SUNY researchers interpret the angle of the

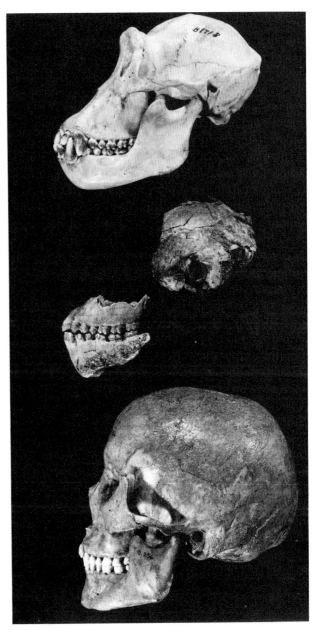

Crania compared: These profiles of a human, chimpanzee and *afarensis* crania show how very apelike the first known hominid was. (Courtesy of the Cleveland Museum of Natural History.)

iliac blade of the pelvis in *A. afarensis* to imply a method of balance during bipedalism more like that of a chimpanzee than a human, that is with a bent hip. They also claim that the lunate articular surface of the socket (the acetabulum) into which the head of the femur fits in the pelvis is less complete in *A. afarensis* than in modern humans. It is incomplete in a region that in humans takes stress when the fully extended hindlimb passes beneath the hip joint. Ergo, this kind of stride does not occur in *A. afarensis*.

Completing the case for a bent-hip, bent-knee walking posture is the suggestion by the SUNY researchers that the knee joint cannot lock in a fully flexed position, as it does in modern humans. The Kent State researchers dispute the description of the anatomy in these three points, and therefore reject the functional interpretation.

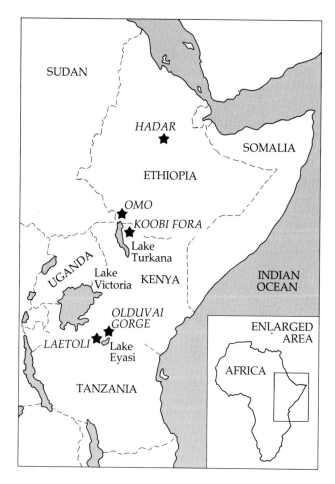

East Africa: The major sites of discovery of *Australopithecus afarensis* fossils, Hadar in Ethiopia and Laetoli in Tanzania, are separated by 1500 kilometers. Evidence of *afarensis* has also been recovered from Koobi Fora in northern Kenya.

The shape of the joint surfaces of certain bones in the foot (the metatarsals) can be taken to imply a greater ability for flexion, which would be useful for climbing, and a poorly developed stability when in a toe-off position. If *A. afarensis* really did employ a bent-hip, bent-knee posture, then it would not have used toe-off anything like the degree that occurs in the modern human striding gait (unit 13).

Finally, Jungers has recently examined the size of hindlimb joints—particularly the femoral head—in modern apes, humans, and *A. afarensis*. The rationale was that if you distribute your body weight on four limbs for most of the time—as chimpanzees and gorillas do for instance—then, relatively speaking, the joint surfaces of the lower limbs would not have to be as extensive as they would if you permanently balanced your full weight on the hindlimbs, as humans do. Sure enough, humans have much larger femoral head surfaces that would an African ape of the same size. And although the femoral head surface in *A. afarensis* is larger than that of an ape of the same size, it is nowhere near the human range. This leads Jungers to conclude that 'the adaptation to terrestrial bipedalism in early hominids was far from complete and not functionally equivalent to the modern human condition'.

One other aspect of the postcranial anatomy worth noting in relation to the biology of *A. afarensis* is the structure of the hands. Although they have often been characterized as 'surprisingly modern', they are in fact rather apelike in manipulative capacity as well as in overall curvature. For instance, the thumb is shorter than in the human hand, and the finger tips are much narrower. Human finger tips are broad, a trait related to the high degree of innervation required for fine manipulative skills. It should be noted that the earliest stone tools so far recognized from the record date to about 2.5 million years, which is post-*afarensis*.

As Johanson and White noted in their early descriptions of the species, *A. afarensis* appears to display extreme sexual dimorphism in body size: the small individuals (females) stood a little over a meter tall and weighed about 30 kilograms, the large individuals (males) 1.7 meters and 65 kilograms. A difference in body size of this degree often implies competition between males for access to females, and this is usually accompanied by large canine size and dimorphism.

Overall, *A. afarensis* anatomy—and presumably

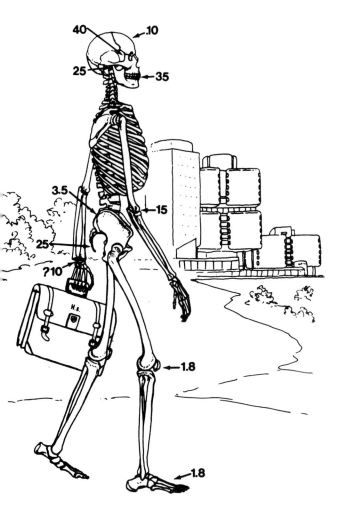

Anatomical heritage: Different parts of our skeleton have been shaped at different periods in our history. The bulbous cranium, for instance, is our most recent acquisition, at 0.10 million years, while the basic dental pattern is the oldest, at 35 million years; other parts fall in between. (Courtesy John Fleagle.)

behavior—is somewhat intermediate between ape and human, a pattern that does not exist today. As a result, there is a tendency to see the adaptation as 'unstable', as being in the process of 'perfection' towards the human model. In fact, this unique anatomical and behavioral repertoire was probably a stable evolutionary package—which, afterall, is

what a biological species is—that lasted several million years. One hint in this respect is that in a species, *A. africanus*, that appears to have closely followed *A. afarensis*, many aspects of the postcranial skeleton are more or less identical to that of *A. afarensis*.

Key questions:

● How might one's interpretation of *A. afarensis* anatomy be influenced if one was (a) most familiar with ape anatomy, and (b) most familiar with human anatomy?
● What might convince one that *A. afarensis* anatomy was indeed 'in transition' and not a stable adaptation?
● How does the knowledge of extant species constrain the interpretation of behavioral scenarios for paleospecies, in general and in *A. afarensis* in particular?
● Would information about the paleoecology of *A. afarensis* help in determining the biology of the species?

Key references:

Andrew Hill, 'Early hominid from Baringo, Kenya', *Nature*, vol 315, pp 222–224 (1985).
Donald C. Johanson and Tim D. White, 'A systematic assessment of early African hominids', *Science*, vol 203, pp 321–330 (1979).
William L. Jungers, 'Relative joint size and hominid locomotor adaptations with implications for the evolution of hominid bipedalism', *Journal of Human Evolution*, vol 17, p 247 (1988).
B. Latimer *et al.*, 'Talocrural joint of African hominoids', *American Journal of Physical Anthropology*, vol 74, pp 155–175 (1987).
Mary Leakey *et al.*, 'Fossil hominids from the Laetolil Beds', *Nature*, vol 262, pp 460–466 (1976).
Randall L. Susman, Jack T. Stern, and William L. Jungers, 'Arboreality and bipedality in the Hadar hominids', *Folia Primatologica*, vol 43, pp 113–156 (1984).
special issue. 'Pliocene hominids fossils from Hadar, Ethiopia', *American Journal of Physical Anthropology*, vol 57, number 4 (1982).

17 / The australopithecines

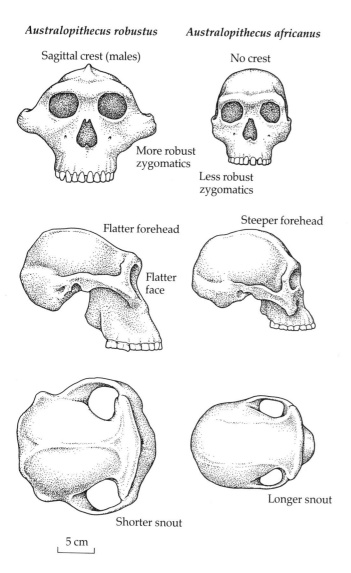

Australopithecus robustus

Sagittal crest (males)

More robust zygomatics

Flatter forehead

Flatter face

Shorter snout

Australopithecus africanus

No crest

Less robust zygomatics

Steeper forehead

Longer snout

5 cm

Cranial comparisons: The principal difference between robust and gracile australopithecines is the dental apparatus: the robust species were apparently eating a diet that required more chewing, as reflected in the large size of the cheek teeth, the increased robusticity of the jaw, and the enlarged attachments for the masticatory muscles. The chewing mechanics dictate the overall shape of the face and cranium. (Courtesy of A. Walker and R.E.F. Leakey/*Scientific American*, 1978, all rights reserved.)

The hominid family began, of course, with a single species, which may have been *Australopithecus afarensis*, close to 5 million years ago. As usually happens with newly established mammalian lineages, this first species gave rise by stages to a range of descendants, producing a relatively luxoriant evolutionary bush. Inevitably, individual branches were pruned away from time to time as species became extinct, and new ones were added. Eventually, and atypically for mammalian groups, the hominid bush was reduced to just one species— *Homo sapiens sapiens*—as the family's sole representative. (Incidentally, horses followed a similar, unusual evolutionary pattern.)

If one were able to go back, say, 2 million years to Africa, one would find several hominid species, perhaps sharing the same habitat, as some species of Old World monkeys do today for instance, or perhaps occupying different habitats, as do modern chimpanzees and gorillas. How many hominid species coexisted on the continent then is a matter of debate and uncertainty: no less than three, perhaps six, maybe more.

In any case, it seems clear that, however many hominid species there were in Africa 2 million years ago, they could be divided into two groups: one with relatively large brains, the other with relatively small brains. The large-brained species were members of the genus *Homo*. So far only one species has been formally named in this group— *Homo habilis*—but many scholars believe that others can be identified. If indeed several species of *Homo* existed at this time, only one of them could have been ancestral to modern man, while the rest became extinct. Labelling of the second group is more contentious, but for simplicity we will call them the australopithecines (members of the genus *Australopithecus*): they all became extinct.

In this unit we will discuss the anatomy and biology of the australopithecines; unit 18 will address the *Homo* group, principally *Homo habilis*; and in unit 19 we will consider current hypotheses on how these various hominid species were related to each other—their family tree, or phylogeny.

Like all early hominids, the australopithecines were essentially bipedal apes with modified dentition. The hominid mode of locomotion and the hominid dental apparatus are likely to have been adaptations to a habitat—and therefore diet—that was different from what is traditionally associated with apes (unit 14). Hominids appear to have lived in a more open environmental setting, not the open

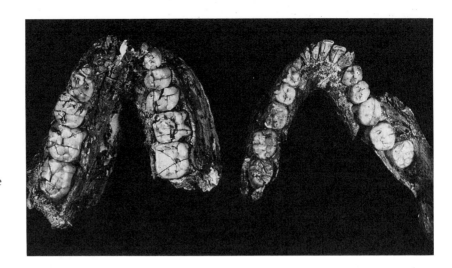

Comparison of lower jaws of *Australopithecus robustus* (left) and A. *africanus* (right): Note the massive molar teeth in the *Australopithecus robustus* mandible from Swartkrans (left) compared with that of A. *africanus* from Sterkfontein (right). (Courtesy of Milford Wolpoff.)

plains of traditional stories, but bushland and woodland savannah. Food was probably in widely scattered patches and, judging from the structure of hominid teeth and jaws, appears to have required more grinding than an ape's diet.

Male australopithecines were significantly larger in body size (about 30 per cent taller, 45 per cent heavier) than females, which probably implies competition among the males for access to the females (unit 11). Australopithecines were social animals, with a social structure not unlike chimpanzees (unit 12).

Judging from a comparison of fossil discoveries from the 2 million-year-old record, australopithecines were about as common on the landscape as other large, open-country primates, specifically baboons. This implies that foraging strategies of hominids and baboons were not *dramatically* different: for instance, had australopithecines been carnivores, their population density would have been much lower than that of the principally-vegetarian baboons. It seems a reasonable conclusion from this observation, and from evidence of wear patterns on australopithecine teeth, that these creatures were principally vegetarian too.

The australopithecines of 2 million years ago occurred in two forms: the 'gracile' (which means slender) and the 'robust' (which is self-evident). So far, only one gracile species of *Australopithecus* has been identified whereas as many as four robust species have been named. The terms 'gracile' and 'robust' seem to imply substantial anatomical differences between the two forms, one small and delicately built, the other bigger and generally more massive. In recent years, however, scholars have come to realize that the difference between the two

forms is mainly in the dental and facial adaptations to chewing: the robust forms have bigger grinding teeth, more robust jaws, and more bulky chewing muscles and muscle attachments.

In the remainder of the skeleton, the gracile and robust australopithecines were roughly comparable, with the robust species having slightly greater stature. Estimates of brain size, which is based on a small number of specimens, typically give the robust species an edge over their gracile cousins, but in fact both are very close to 500 cm^3 (unit 27).

No undisputed australopithecine fossil has been found outside the African continent. In fact, most scholars agree that hominids did not leave Africa before about 1 million years ago, when *Homo erectus* expanded its range to include Eurasia (unit 21).

We will now look more closely at some of the australopithecine anatomy that qualifies them as hominids, and some of the differences between the gracile and robust forms. This survey will include the teeth, jaws and cranium, the pelvis and associated locomotor skeleton, and finally the hands. In each case we will discuss the functional implications of the anatomy.

The teeth, jaw, and cranial anatomy is really one functional complex. As we saw in unit 15, the hominid dental adaptation can be described in general as moving in the direction of producing a grinding machine. And the difference between the two forms of australopithecine is that the robust species have taken this adaptation to an extreme, having enormous, flat molars, and relatively small, blade-like incisors and canines.

This exaggeration of the hominid dental adaptation is seen in several other respects in the robust australopithecine group too. For instance, in all hominids, the tooth row is tucked under the face more than in apes, giving them a less projecting facial profile and increasing chewing efficiency. In the robust australopithecines this tucking under is particularly marked. The robusticity of the lower jaw (mandible) that is characteristic of hominids compared with apes is particularly apparent in the robust species, reflecting the more powerful chewing action.

The extra muscle power necessary for this chewing action in the robusts has two anatomical consequences. First, one of the muscles that powers the lower jaw—the temporal muscle—is anchored to a raised bony crest that runs along the top of the cranium, front to back. This so-called saggital crest, which is also present in gorillas, is absent in gracile australopithecines. Second, the great size of the temporal muscle in robust australopithecines, and a second chewing muscle, the masseter, causes the cheek bones (the zygomatic arch) to be exaggerated and flared forward. This, and the strengthening of the central part of the face by pillars of bone, gives the robust australopithecine face a characteristic 'dished' appearance.

The very obvious difference in dental apparatus between gracile and robust australopithecines was once interpreted as a the result of a substantial difference in diet: the robust species were hypothesized to be principally vegetarian, the gracile species

much more carnivorous. In 1980, however, Alan Walker of Johns Hopkins University analyzed the microwear patterns on the tooth surfaces using the scanning electron microscope, and found that both types of australopithecine conformed to a vegetarian pattern. More specifically, the microwear was rather like that of chimpanzee or orangutan, which eat various forms of fruit.

Walker therefore suggested that the robust australopithecine dental apparatus was more exaggerated than its gracile cousin's, not because it necessarily ate different food, but because it ate more of the same type of food: being a bigger animal, the robust species needed more fuel. More recently, however, Frederick Grine of the State University of New York at Stony Brook and Richard Kay of Duke University, performed more tests with the scanning electron microscope, and came to slightly different conclusions.

Yes, they agreed that the two australopithecine forms were basically vegetarian, as Walker had said. But they concluded that the robust species consumed foods that were tougher than those eaten by the gracile species. The difference, they suggested, was rather like that between the spider monkey, which eats fleshy fruits, and the bearded saki, which lives on seeds encased in a tough covering. Such a dietary difference is consistent with

Comparison of gracile and robust australopithecine skulls: The skull of a gracile australopithecine, *Australopithecus africanus*, from Sterkfontein, South Africa (left) and the skull of a robust australopithecine, from Lake Turkana, Kenya (right). Note the more massive face, saggital crest and pronounced zygomatic arches in boisei: these are all anatomical adaptations to massive musculature that powered the larger mandible in this large australopithecine species. (Courtesy of Peter Kain and Richard Leakey.)

evidence that robust australopithecines lived in drier habitats, where soft fruits and leaves would be uncommon.

In terms of function and overall size, the post-cranial skeletons (that is, from the neck down) of gracile and robust australopithecines are very similar to each other, as far as can be deduced from the limited amount of fossil material available. Estimates of body size are tentative, because of the fragmentary nature of the fossil record and the uncertainties of extrapolating from experience with living animals to extinct creatures.

Using a model based on apes but not humans, William Jungers of the State University of New York at Stony Brook has recently produced the following figures. For *Australopithecus africanus* the mean body weight (for males and females) is 52.9 kilograms, with a range from smallest to largest of 33.3 to 67.5. The figures for *Australopithecus robustus* are a mean of 62.0 kilograms, and a range of 42.2 to 88.6. The mean for *Australopithecus boisei* is 58.8 kilograms, and a range of 36.9 to 88.6. For comparison the mean weight of *Australopithecus afarensis* is 57.8 kilograms and of *Homo sapiens sapiens* is 60.0 kilograms. These figures are in very close agreement with those produced independently by Henry Mettenry of the University of California.

The australopithecine pelvis of 2 million years ago was very much like Lucy's a million years earlier, which we described in unit 14 as being adapted to upright walking rather than quadrupedalism. The thigh bone is different from the typical *Homo* pattern: the head of the femur is smaller than in *Homo* and is attached to a longer, more slender neck. Once interpreted as implying less efficient bipedalism than in *Homo*, this pattern is now considered simply to reflect a different style of bipedalism, with possibly even greater biomechanical efficiency. So far no complete foot has been unequivocally identified as gracile or robust australopithecine. However, individual bones from South African sites indicate that, although these species were fully bipedal, they were as adapted to climbing as was *Australopithecus afarensis* (unit 14).

There was a difference, however, between Lucy's hands and those of the later australopithecines. The hand bones of *Australopithecus afarensis* were strikingly apelike—having curved phalanges, thin tips to the fingers, and a short thumb. By contrast, recent analysis of robust australopithecine hand bones from the site of Swartkrans indicates that they were much more humanlike. Randall Susman of the

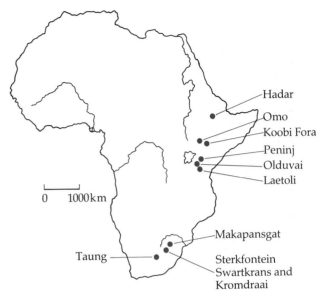

Australopithecine sites: Australopithecine fossils have been discovered only in Africa, specifically in East and South Africa. The first discovery was at Taung, in South Africa, with the other South African sites, Sterkfontein, Swartkrans, and Kromdraai following soon after; later still, in the 1950s, Makapansgaat was excavated. With the 1959 discovery of *Zinjanthropus* at Olduvai Gorge in Tanzania, the emphasis of fossil finds began to turn to East Africa. In addition to Olduvai, the major sites are Koobi Fora (Kenya), the Lower Omo Valley (Ethiopia), and then the Hadar (Ethiopia).

State University of New York at Stony Brook reports that the thumb is longer and more mobile, and that the fingertips were much broader, a feature thought to be associated with the supply of blood vessels and nerve endings to the sensitive finger pads.

Susman considers that, anatomically, the robust australopithecines probably possessed sufficient manipulative skills to be able to make stone tools, an ability that has usually been thought of as strictly within the domain of *Homo*. Indeed, C.K. Brain has recently identified what he takes to be bone tools—digging sticks—associated with robust australopithecine fossils from Swartkrans. If the robust species were tool makers then the likelihood is that the gracile species was too.

The gracile and robust australopithecines have often been viewed as basically the same animal, but built on different scales. Functionally speaking, this notion is true in many respects. But the relationship was also viewed in terms of evolutionary progression, that the gracile species was ancestral to the robust species, in whom the australopithecine traits had become extremely exaggerated: specifically, the chewing apparatus became increasingly

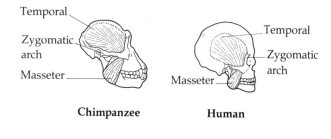

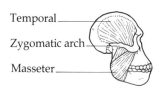

Australopithecus

Anatomy of chewing: Two muscles are important in moving the lower jaw during chewing: the masseter, which is attached to the zygomatic arch (cheek bone), and the temporal, which passes through the arch. The bigger the masseter and temporal muscles, the bigger the zygomatic arch. Chimpanzees have about three times as much chewing-muscle bulk than humans, and the australopithecines even more.

robust. If this were true, then the fossil record should have revealed a steady increase through time in dental, facial, and jaw robusticity.

A 1985 discovery from the west side of Lake Turkana finally put to rest this simple relationship. The discovery, made by Alan Walker, was of an australopithecine cranium as robust as any yet known, and yet it was 2.5 million years old. Clearly, the huge molars, flared cheek bones, and dished face could not be the end product of an evolutionary line if it were already present at the origin of that supposed line. What this discovery does to the shape of the hominid family tree is still in discussion, which we will address in unit 19.

Walker's discovery, known as the black skull, was surprising not only for its great age but also

for its unexpected combination of anatomical characteristics. For, although the face was distinctly like that of the most robust of robust australopithecines, *Australopithecus boisei*, the cranium—particularly the top and back—were not: they were just like that of the most primitive known hominid, *Australopithecus afarensis*. Such a combination of anatomical packages was a surprise to most people, and a reminder than hominid biology of 2 million years ago was more complicated than current hypotheses have allowed.

Key questions:

● Under what circumstances might the original hominid species have evolutionarily diverged, producing a diversity of descendants that occupied different habitats?
● How clearly would this habitat separation be reflected in the fossil record?
● In what ways did the biology of australoplithecines 2 million years ago differ from that of their ancestor, *Australopithecus afarensis*?
● Why might the biomechanics of australopithecine bipedalism have differed from the *Homo* model?

Key references:

Eric Delson, 'The evolution and paleobiology of robust *Australopithecus*', *Nature*, vol 327, pp 654–655 (1987).

Frederick E. Grine (editor), *Evolutionary history of the robust australopithecines*, Aldine, 1989.

Frederick E. Grine and Richard F. Kay, 'Early hominid diets from quantitative image analysis of dental microwear', *Nature*, vol 333, pp 765–768 (1988).

Roger Lewin, 'The first ape-men', in *In the age of mankind*, Smithsonian Books, 1988.

Alan Walker *et al.*, '2.5–Myr *Australopithecus boisei* from west of Lake Turkana, Kenya', *Nature*, vol 322, pp 517–522 (1986).

18 / Early *Homo*

Two distinct groups of hominids coexisted 2 million years ago, in Africa. One group was made up of small-brained species, the australopithecines (unit 17); in the second included the large-brained hominids, members of the genus *Homo*. In this unit we will discuss what is known of the anatomy and biology of early *Homo* and consider some of the uncertainties that surround the origin and composition of this group.

The definition of the genus *Homo* has always been somewhat contentious, not least because it is tied—consciously or unconsciously—to the state of 'being human'. There is a series of anatomical characters that are to be found uniquely in *Homo*—some aspects of the pelvis, thigh bones, teeth, face and cranium, for instance—but what has always stood out prominently in scholars' definitions is the size of the brain. To be *Homo* is to be a large-brained hominid, one presumably more technologically accomplished than the australopithecines. The question is, how big a brain qualifies for admission to the genus *Homo*?

The average modern human brain is 1350 cm^3 in capacity, with a range of 1000 cm^3 to about 2000 cm^3. How much smaller than 1000 cm^3 can a hominid's brain be, and still be counted as *Homo*? Before 1964 there were several estimates of this 'cerebral rubicon', ranging from 700 to 800 cm^3. In the late 1940s the British anthropologist Sir Arthur Keith proposed a figure of 750 cm^3, which lies midway between the largest known gorilla brain and the smallest human brain. Keith's proposal was widely accepted until 1964, when Louis Leakey, Phillip Tobias and John Napier advanced a new definition of the genus *Homo*, which included a cerebral rubicon of 600 cm^3.

Leakey and his colleagues' new definition was associated with the announcement of a new species of *Homo*—*Homo habilis*—fossils of which had been recovered from Olduvai Gorge between 1960 and 1963. The reason for the reduction in the cerebral rubicon in this new definition of *Homo* was that the fossil cranium that was part of the new species had a capacity of only 680 cm^3, a figure that would have failed under Keith's standard.

Leakey's fossils came from the same site where *Zinjanthropus* had been discovered, and were therefore of the same age, 1.75 million years old. *Homo habilis* was therefore the earliest member of the *Homo* lineage to be recognized, and Leakey considered it to be directly ancestral to *Homo sapiens*, *Homo erectus* being a sidebranch, not an intermediate species. By contrast, most other scholars see a continuous evolutionary line, going from *Homo habilis*, through *Homo erectus*, to *Homo sapiens*.

The announcement of *Homo habilis* occasioned tremendous objections from the anthropological community, some of which are still voiced today. In an article published in 1986, British anthropologist Christopher Stringer explained the substance of the negative reaction as follows: 'The two main arguments against the existence of *Homo habilis* have centered on the supposed lack of 'morphological space' between *Australopithecus africanus* and *Homo erectus* for such a species, and the sheer variation in specimens assigned to the species.' The title of Stringer's article, 'The credibi-

***Homo habilis* skull KNMER 1470:** Reconstructed from many fragments found on the eastern shore of Lake Turkana, Kenya. Note the less protruding face and rounder, larger cranium than seen in the australopithecines. 1470 has a cranial capacity of about 750 cm^3 and is dated at 1.9 million years. Courtesy of Peter Kain and Richard Leakey.

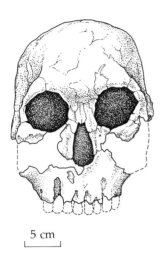 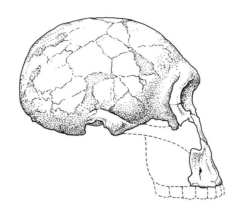 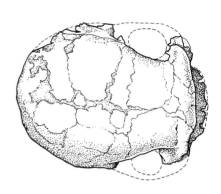

5 cm

Three views of the *Homo habilis* skull KNMER 1470 from Koobi Fora in Kenya: Note the rounder cranium compared with the australopithecines. The postorbital constriction (the distance between the temples) is less pronounced here due to brain enlargement, about 750 cm³. (Courtesy of A. Walker and R.E.F. Leakey/ *Scientific American*, 1978, all rights reserved.)

lity of *Homo habilis'*, gives some idea of the depth of current uncertainties surrounding the putative species *Homo habilis*.

The first of the two arguments—the question of morphological space—is based on the proposal that *Australopithecus africanus* is ancestral to *Homo erectus*. The argument is that, although there are some clear anatomical differences between these two species—notably a larger brain in the latter—they are in fact very similar in many ways. They are *so* similar, the argument runs, that it is hard to imagine something that is intermediate between the two that could not readily be described as being a normal variant of either the putative ancestor or the putative descendant. (The phrase 'morphological space' simply refers to the extent of anatomical difference between the two species.) Indeed, at least half of the specimens that have been assigned to *Homo habilis* by some scholars have been variously attributed to *Australopithecus africanus* or *Homo erectus* by others.

The second argument centers on the range of anatomical variation among the specimens that have variously been said to belong to *Homo habilis*. This variation has to do with cranial and dental shape as well as brain size, which ranges from close to 600 cm³ to more than 800 cm³. Interpretations of this anatomical variability differ. Some scholars say that the range of anatomical differences within this group of fossils is nothing more than normal variability within a single species, including differences between large males and small females.

Others follow the line of the morphological space objection, and say there is no such species as *Homo habilis*. Still others consider that the fossils represent two species, one of which is *Homo habilis* while the other is a second species of *Homo*, which so far has not been formally named. No wonder that Stringer titled his article as he did.

We will now cover briefly some of the anatomy that separates early *Homo* from the contemporary australopithecines, and consider how this might be related to new behavioral adaptations. Finally we will discuss the discovery of a new fossil that has been attributed to *Homo habilis* and explore how it throws further light on the biology of this species and on the tempo of human evolution as a whole.

First, as we have already noted, the brain capacity is larger than in australopithecines, a change that produces several associated characteristics. For instance, the temple areas in australopithecines narrows markedly (best seen from top view), forming what is known as the postorbital constriction. In early *Homo* this constriction is much reduced because of the expanded brain. In addition, in australopithecines the face is large relative to the size of the cranial vault, a ratio that is reduced in the larger-brained *Homo* species. The cranial bone itself is thinner in *Homo* than in *Australopithecus*.

The tooth rows are tucked under the face as in other early hominids, a feature that becomes even more exaggerated in later species of *Homo*. The jaw and dentition of *Homo habilis*, however, is less massive than in the australopithecines. Although the teeth are capped with a thick layer of enamel, their overall appearance is less of a grinding machine than appears in the small-brained hominids: the cheek teeth are smaller, and the front teeth larger than in australopithecines, and the premolars are narrower. The patterns of wear on *Homo habilis* teeth are, however, indistinguishable from those of the australopithecines: the pattern is that of a generalized fruit eater. Only with the evolution of *Homo erectus* 1.6 million years ago does the tooth-wear pattern make a dramatic shift, a change that might indicate the inclusion of a significant amount of meat in the diet.

The original set of *Homo habilis* fossils from Olduvai Gorge included a relatively complete hand, the structure of which was compatible with an ability to make and use tools, concluded John Napier. Indeed, the name *Homo habilis* means handy man. Stone artifacts—'choppers', 'scrapers', and sharp flakes of the Oldowan technology (unit 24)—first appear in the record about 2.5 million years ago, which, according to some estimates, coincides with the evolution of *Homo habilis*. Such a coincidence falls in neatly with the long-established tendency to view stone-tool making as a uniquely human—that is, *Homo*—ability. But, as we saw in the previous unit, both the gracile and robust australopithecines may also have had the ability to make and use stone artifacts.

The evolution of technological skills associated with stone-tool making has always appeared to be a satisfactory explanation for the expansion of brain capacity in the *Homo* lineage. If australopithecines were in fact equally skillful, then this explanation falls. There must, presumably, have been some selection pressure on mental skills that separated the *Homo* and australopithecine lineages. Whether this was associated with the development of more complex subsistence activities, or was instead in the realm of more complex social interaction (unit 27), is difficult to determine.

Whatever was the behavioral complex that included a demand for greater brain power, *Homo habilis* apparently retained the tree-climbing ability that was evident in the anatomy of the earlier hominid, *Australopithecus afarensis*. In 1982 Randall Susman and Jack Stern of the State University of New York at Stony Brook published an analysis of the foot and hand bones that were part of that initial *Homo habilis* discovery at Olduvai Gorge. 'The skeleton represents a mosaic of primitive and

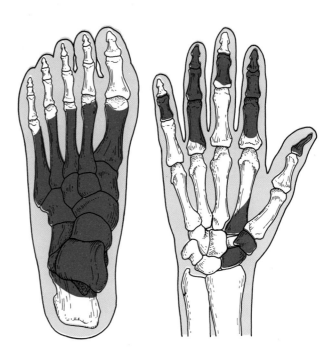

Reconstruction of the hand and foot fossil bones (shaded) of *Homo habilis* from Olduvai Gorge, Tanzania: The foot has all the appearance of modern form, though, because of their absence, nothing can be said of the toes. The hand is essentially modern too, but the curvature in the finger bones and the heavy point for muscle attachment are both evidence of a powerful grip, possibly used in frequent arboreality. The vasculature and nerve supply to the finger tips is greater in *Homo habilis* than in earlier hominids, indicating a greater sensitivity and manipulative skill.

derived features, indicating an early hominid which walked bipedally and could fabricate stone tools but also retained the generalized hominoid capacity to climb trees', they said. This conclusion is questioned by some scholars, as too is the contention that the foot and hand belonged to the same, 13-year-old, individual.

In absolute terms, fossil hominid collections are fairly sparse, in spite of the several spectacular finds of 'partial skeletons'—of *Australopithecus afarensis* and *Homo erectus*—in recent years. Nevertheless, it has seemed fairly clear that in the earliest hominid, and in australopithecines in general, there was a considerable difference in body size between males and females of each species. It has also seemed clear that in *Homo erectus*, which evolved about 1.6 million years ago, this marked sexual dimorphism is greatly reduced, largely by an increase in the size of the female. And it has often been assumed that the first member of the *Homo* lineage would be somewhat intermediate between these two states.

The discovery by Donald Johanson and his

colleagues in 1986 of a highly fragmented partial skeleton of *Homo habilis* from Olduvai Gorge effectively abolished this perception. In fact, as British anthropologist Bernard Wood commented at the time of publication of the fossil: 'The new find rudely exposes how little we know about the early evolution of *Homo*.'

The 1.8 million-year-old fossil—some 300-plus fragments of it—was that of a mature female, and yet she stood only a meter tall, slightly less than the diminutive Lucy. Her arms were very long: the ratio of arm length to leg length was 95 per cent—a little higher than *Australopithecus afarensis*—as compared with 70 per cent in modern humans. Although the fossil, code-named OH 62, is very similar in many ways to *Australopithecus afarensis*, its discoverers argue that it is *Homo habilis*, on the basis of comparison with a specimen from Sterkfontein in South Africa. The cranium of OH 62 is too fragmented and incomplete to allow a reliable estimate of brain size, but the assumption is that it would have been considerably larger than Lucy's.

The discovery of OH 62 shows that sexual dimorphism is just as marked in *Homo habilis* as it is in earlier and contemporary australopithecines: the species is not neatly intermediate between a 'primitive' australopithecine ancestor and an 'advanced' *Homo erectus* descendant. 'The very small body size of the OH 62 individual suggests that views of human evolution positing incremental body size increase through time may be rooted in gradualistic preconceptions rather than fact', note Johanson and his colleagues in their 1987 publication. 'This reinforces the view that encephalization in the terminal Pliocene played a key role in hominid evolution.'

It should be noted that not all anthropologists agree that OH 62 is a member of *Homo habilis*. Indeed, some take it as evidence that there were two *Homo* species in Africa 2 million years ago, not one. The debate continues.

If OH 62 is *Homo habilis*, and if, as most people agree, this species gave rise to *Homo erectus* (but see unit 21), then the evolutionary transition must have been abrupt. The time gap between OH 62 (dated at 1.8 million years) and the earliest *Homo erectus* (dated at 1.6 million years) is relatively small.

Fossils attributed to *Homo habilis* have been recovered from various sites in South and East Africa, and include the famous 1470 man from Koobi Fora, in Kenya. The oldest specimen dates to a little more than 2 million years, and the youngest

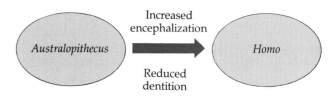

Hominid trends: The transition between *Australopithecus* and *Homo* was accompanied by an increase in brain size and a decrease in the robusticity of the cheek teeth. This trend continued with later species of *Homo*.

about 1.8 million years. Even if the species originated somewhat earlier than this, say 2.5 million years ago, it apparently was still relatively short lived. Although there are some fragmentary indications of a perhaps earlier origin of the *Homo* lineage, a date of 2.5 million years coincides with a major climatic event that engendered speciation in a wide variety of organisms (unit 5). The robust australopithecine lineage also seems to have its origin at this event. The habitat fragmentation generated by the climatic event therefore appears to have sundered the hominid group in several different adaptive directions.

Key questions:

● Apart from direct anatomical information, what other evidence might help determine that early *Homo* pursued different subsistence strategies from the australopithecines?
● Is it more or less likely that there could have been several coexisting species of *Homo* than several coexisting species of *Australopithecus*.
● How could one resolve the question of who made the tools?
● Are there special reasons why the origin of *Homo erectus* might have been an abrupt event?

Key references:

Donald C. Johanson *et al.*, 'New partial skeleton of *Homo habilis* from Olduvai Gorge, Tanzania', *Nature*, vol 327, pp 205–209 (1987).
D.E. Lieberman *et al.*, 'A comparison of KNM-ER 1470 and KNM-ER 1813', *Journal of Human Evolution*, vol 17, pp 503–512 (1988).
Christopher B. Stringer, 'The credibility of *Homo habilis*', in *Major topics in primate and human evolution*, edited by B. Wood, L. Martin, and P. Andrews, 1986.
Randall L. Susman and Jack T. Stern, 'Functional morphology of *Homo habilis*', *Science*, vol 217, pp 931–934 (1982).
Bernard Wood, 'Who is the 'real' *Homo habilis*?', *Nature*, vol 327, pp 187–188 (1987).

19 / Early hominid overview

This unit will explore recent developments and current thinking about how early hominids were evolutionarily related to each other. This subject—phylogeny—has always attracted a lot of attention among anthropologists, often overshadowing the more basic questions of hominid biology, such as subsistence strategies and behavior.

Evolutionary trees—phylogenies—can be constructed only with those species that are recognized. This truism is worth stating because there is good reason to believe that the group of hominid species currently recognized by anthropologists is incomplete, perhaps substantially so. Not only is there a strong possibility that new fossil hominid species will be discovered within the next decade, but also several scholars argue that some existing fossils represent species other than those already formally named. This probable incompleteness of the fossil record should therefore be borne in mind whenever a diagram is offered as an hypothesis of 'the' hominid phylogeny.

Making the assumption that primate species on average have a longevity of about 1 million years, Robert Martin of the Anthropological Institute in Zurich recently calculated that about 6000 primate species have ever existed. The 183 living primate species therefore represent just 3 per cent of this total. More particularly, although 84 species of hominoid are estimated to have existed during the past 35 million years, only about half are known from the fossil record. And of the 16 hominid species that are calculated to have existed in the past 8 million years or so, a maximum of nine have been named among the fossils recovered to date. The family trees that are constructed today therefore probably have half their branches missing.

As noted earlier, some of those branches may be occupied when new fossil species are recovered from the ground. But, as Ian Tattersall has recently argued, other factors may also apply. 'Over the past several years increasing attention has been paid to the search for patterns in the human fossil record', notes Tattersall, an anthropologist at the American Museum of Natural History. 'The reliability of any attempt to recognize pattern, however, is constrained by the accuracy with which we are able to recognize species in that record....[It] is hard to avoid the conclusion that under current taxonomic practice there is a distinct tendency to underestimate the abundance of species in the primate, and notably the hominid, fossil record.'

The taxonomic practice to which Tattersall refers has to do with the interpretation of anatomical variability. During the first half of this century it was common for scholars to apply a new species name to virtually each new fossil unearthed. Each variant in anatomical structure was taken as indicating a separate species, an approach known as splitting. The result was a plethora of names in the hominoid record. In 1965 Elwyn Simons and David Pilbeam, both then at Yale University, rationalized this paleontological mess, and reduced the number of genera and species to a mere handful. This approach, in which anatomical differences are

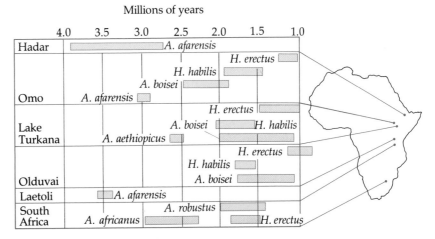

Millions of years

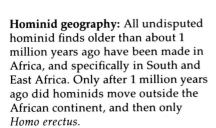

Hominid geography: All undisputed hominid finds older than about 1 million years ago have been made in Africa, and specifically in South and East Africa. Only after 1 million years ago did hominids move outside the African continent, and then only *Homo erectus*.

Most likely evolutionary pattern

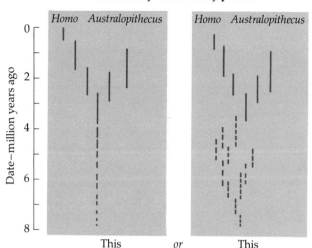

This   *or*   This

Possible trees: Hominid phylogenies are typically drawn with a single stem beginning between 3 and 4 million years ago. If, however, the origin of the hominid lineage was somewhere between 5 and 8 million years ago, as indicated by molecular evidence, then thought has to be given to the shape of the tree prior to 4 million years ago. Is it likely to be a single stem right back to the beginning? Or is it possible that earlier radiations existed, of which the hominids known between 3 and 4 million years ago are merely a recent radiation?

interpreted as variation within species rather than variation between species, is known as 'lumping'.

Lumping became the guiding ethic of anthropology, and in the extreme manifested itself as the 'single species hypothesis', which was popular during the 1960s and early 1970s. Promulgated by Loring Brace and Milford Wolpoff of the University of Michigan, this notion explained all anatomical differences among hominid fossils at any point in time as within-species variation. In other words, only one hominid species existed at any one time, making the hominid family tree a progression of species through time, each following on from the one before.

Although the single species hypothesis is no longer considered valid, there is, according to Tattersall, still a tendency to interpret anatomical differences as within-species variation rather than among-species variation. One reason is that, because of the nature of the system, there is no practical guide for how much anatomical difference between two fossils signals the existence of separate species. 'The reason for this is, of course, that there is no direct relationship, indeed no consistent relationship at all, between speciation and morphological change', says Tattersall.

In other words, in some cases a daughter species

might diverge from the parental species with very little obvious anatomical difference developing in the process, while in other cases considerable differences might occur. Unless you have the living animals in front of you and can observe their behavior, it is often impossible to know whether the individuals belong to one species or two. Given this, it is obviously easier to subsume anatomical differences under within-species variation rather than try to argue for separate species. And this has certainly become a tradition in anthropology. The result, argues Tattersall, 'is simply to blind oneself to the complex realities of phylogeny'. In other words, the true hominid family tree—the one that actually happened in evolutionary history—almost certainly is more bushy than the ones currently drawn by anthropologists.

Although most anthropologists would regard Tattersall's position as somewhat extreme, many are coming to accept that hominid phylogeny is indeed more complex than it is usually portrayed. And this was emphasized by the rethinking that was provoked by the 1985 discovery of the black

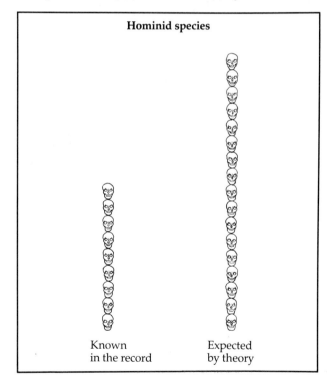

Hominid species

Known in the record

Expected by theory

More fossil hominids?: Theoretical calculations based on what is known of other primates indicate that there should have been about 16 hominid species between 7.5 million years ago and the present. So far only nine (at maximum) have been identified. Are others yet to be identified? Or were hominids different from other primates in their speciation and extinction patterns?

skull, a robust australopithecine that did not immediately fit into the prevailing phylogenetic picture (unit 17).

As we have already seen, the hominid lineage probably evolved between 8 and 5 million years ago, and in this unit we are concerned with its evolutionary fate through to 1 million years ago.

The first point to make is that the earliest known species is *Australopithecus afarensis*, remains of which have been dated at between 3.7 and 2.8 million years ago. All other known hominid species are dated later than 3 million years ago. It is of course possible that *afarensis* was indeed the first member of the hominid lineage, and that no other species evolved until after 3 million years ago. But, given the typical evolutionary pattern of mammalian groups, this seems unlikely. The typical evolutionary pattern is that, once a new lineage is established there follows a quite rapid radiation of species: in other words, the group, or clade,

is bushy from the beginning, not just half way through.

It therefore remains to be seen whether *A. afarensis* is itself the tip of an existing hominid bush, other branches having become extinct. At the moment, however, this must remain speculative. Moreover, the fossil gap between 3.7 and 5 to 8 million years ago will be difficult to fill, because in much of their anatomy—especially their dental and cranial anatomy—these earliest hominid species will be very difficult to distinguish from the earliest ancestors of the modern African apes. The hominids will, however, have been bipedal, the distinguishing characteristic of the clade.

A survey of recent books and papers in physical anthropology reveals a rather varied collection of proposed hominid phylogenies. This diversity of professional opinion is instructive, confirming that 'hominid phylogeny is still far from being resolved', as two British scholars recently put it. There

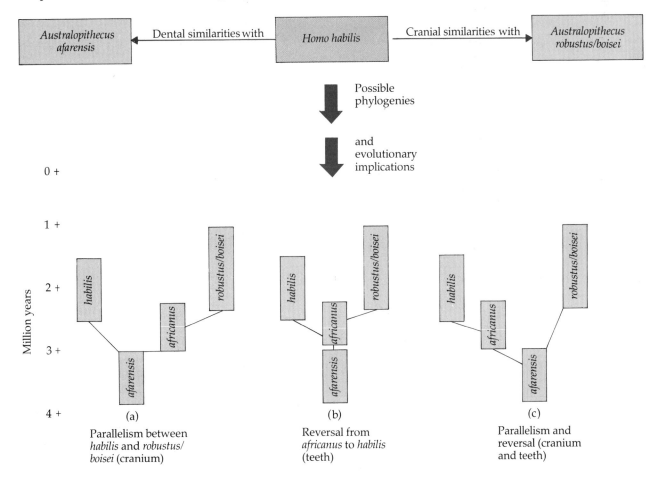

Parallelism and other complexities in hominid history: Phylogenetic trees often involve complex evolutionary, and hominid history is no exception. The anatomical similarities among the early hominids means that any proposed phylogeny will involve parallelism (a), reversal (b), or both (c).

is, however, rather broad—but not universal—agreement that all the known hominid species from 3 million years onwards trace their origins back through *A. afarensis*, the oldest of all the known hominid species. In other words, the great majority of anthropologists view *A. afarensis* as the stock from which all later hominids derived. Current debate centers on the exact course of these derivations.

This broad agreement has emerged only recently, the result of extensive analysis of the *A. afarensis* material itself and some subsequent discoveries of other fossils. When in the late 1970s Donald Johanson and Timothy White first described the *A. afarensis* fossils and proposed that the species be regarded as the last common ancestor of all later hominids, their scheme was not widely accepted. Many researchers considered that the degree of anatomical and size variation among the fossils from the Hadar and Laetoli indicated the presence of two species, not one.

According to this alternative interpretation, the smaller of the species was a primitive *Australopithecus*, ancestor of the later gracile and robust australopithecines. The second species was said to be a primitive *Homo*, forerunner of *Homo habilis*. A common ancestor between *Homo* and *Australopithecus* would therefore have to be sought earlier in

time than *A. afarensis*, that is earlier than 3.7 million years ago. Most prominent among scholars advancing this argument was Richard Leakey, who still considers it a possibility.

As anthropologists have gradually become persuaded of the core of Johanson and White's original suggestion—that all later hominids derive in some way from *A. afarensis*—debate about hominid phylogeny has come to focus on two key issues: the origin of the *Homo* lineage and the origin of the robust australopithecines.

One clear trend through time in the hominid lineage was the increase in the size of cheek teeth and reduction in size of the front teeth, particularly the canines: this is exemplified in the series, *A. afarensis* to *A. africanus* to *A. robustus/boisei*. As noted in unit 17, this trend appears to have been associated with increased grinding activity as a way of processing food. When one adds *Homo habilis* to the sequence, however, the trend is bucked: although this species comes after *A. africanus* in time, the emphasis on grinding has apparently been reversed. In this respect *Homo habilis* is more like *A. afarensis* than *A. africanus*, a comparison that has to be accounted for by any proposed phylogeny. There were been two main proposals here, both of which would have certain difficulties.

The first proposal viewed *A. afarensis* as the direct

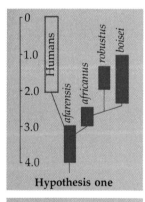

Hypothesis one

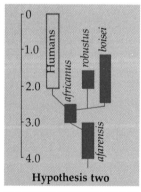

Hypothesis two

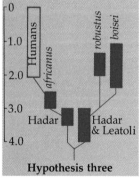

Hypothesis three

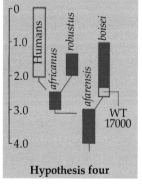

Hypothesis four

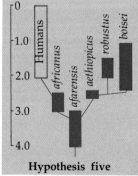

Hypothesis five

A forest of hominid trees: During the past decade a variety of possible hominid phylogenies have been proposed, but the discovery of the black skull has had a major impact on current thinking. Shown here is a selection of proposed trees: some have now been abandoned by their authors or are now out of date (hypotheses one to three); the last two—hypotheses four and five—are now the most strongly supported. Hypothesis four shows the black skull as part of *Australopithecus boisei*, with *A. africanus* giving rise to *A. robustus* and to *Homo* via an unspecified intermediate. Hypothesis five shows the black skull as a separate species, *A. aethiopicus*, which is the sister species (and perhaps ancestor) to the other robust australopithecines. *Australopithecus africanus* is shown as being ancestral to *Homo*. Of these two hypotheses, five is currently the most strongly supported.

ancestor of *Homo* on one hand and *A. africanus* on the other, which is then ancestral to the robust australopithecines (hypothesis one in the diagram). In this scheme the dental similarities between *Homo habilis* and *A. afarensis* are simply explained: the former is the direct descendant of the latter, and therefore carries many anatomical similarities. However, *Homo habilis* also shares some anatomical features with the robust australopithecines, particularly in the cranium. If *Homo habilis* truly is a direct descendant of *A. afarensis*, splitting away before the *A. africanus/A. robustus* lineage had begun to develop, the cranial features in common between *Homo habilis* and the robust australopithecines must have evolved independently and in parallel.

The second proposal viewed *A. africanus* as the last common ancestor between *Homo* and the australopithecines, *A. africanus* having derived directly from *A. afarensis* (hypothesis two). In this case the similarities between *Homo habilis* and the robust australopithecines is easily accounted for by their common heritage through *A. africanus*. However, the dental characteristics shared between *Homo habilis* and *A. afarensis* have to be explained as a reversal of the direction of evolution already established in the hominid lineage: if *Homo habilis* derived from *A. africanus* its dental apparatus would have had to have become reduced in size.

If either of these two schemes were correct—and there still remains the possibility that *Homo habilis* descended from a yet unrecognized species—then it inevitably would involve either parallel evolution or reversal. Although both of these processes are known to occur in evolution, they represent special cases. Their incorporation into an explanatory scheme therefore weakens it to some degree.

Although proposal two—with *A. africanus* as the last common ancestor—was popular for some years, just recently it has been overtaken by events, specifically the discovery in 1985 of an unexpected combination of anatomical features in a 2.5 million-year-old robust australopithecine—the black skull from Koobi Fora in Kenya (see unit 17).

Part of the structure of proposal two was a continual progression of dental robusticity through time in the australopithecines, from *A. africanus* to *A. robustus* to *A. boisei*. However, the discovery of the black skull ruled out the notion of a simple progression, because although it was very close to the beginning of the lineage in time it was already extremely robust. Moreover, the new fossil had

striking similarities in the rear of the cranium with *A. afarensis*. There is therefore a reasonable argument to be made that the new fossil, which some people call *A. aethiopicus*, evolved directly from *A. afarensis*, and was then ancestral to the other robust australopithecine species (hypothesis five). Inevitably, this pushes *A. africanus* to one side: *A. africanus* would remain as the most likely ancestor of *Homo habilis*, but would no longer be ancestral to the robust australopithecines. A phylogeny of this sort is the one currently most favored.

It should be remembered, of course, that although this discussion has focused in part on the origin of *Homo habilis* as a major branch of the hominid phylogeny, several scholars believe there to have been at least two contemporary *Homo* species between 2.5 and 1.6 million years ago. If there is agreement about this point, then published hominid phylogenies will become yet bushier, probably as a prelude to a recognition of yet further complexity.

There is only one true phylogeny—the one that actually happened—and the difficulties encountered in tracing it reveal the complexities of the evolutionary process.

Key questions:

- What type of species would be suitable comparisons for the evolutionary pattern experienced by hominids?
- How would one test the hypothesis that, for instance, *Homo habilis* derived from *A. africanus* rather than from *A. afarensis*?
- What would be the significance of the coexistence of more than one species of *Homo* between 2.5 and 1.6 million years ago?
- What kind of fossil discovery would most upset current views of hominid phylogeny?

Key references:

A.T. Chamberlain and Bernard A. Wood, 'Early hominid phylogeny', *Journal of Human Evolution*, vol 16, pp 119–133 (1987).
Eric Delson, 'Evolution and palaeobiology of robust *Australopithecus*', *Nature* vol 327, pp 654–655 (1987).
Randall R. Skelton *et al.*, 'Phylogenetic analysis of early hominids', *Current Anthropology*, vol 27, pp 21–43 (1986).
Ian Tattersall, 'Species recognition in human paleontology', *Journal of Human Evolution*, vol 15, pp 165–175 (1986).
Timothy D. White *et al.*, '*Australopithecus africanus*: its phyletic position reconsidered', *South African Journal of Science*, vol 77, pp 445–470 (1981).

20 / Hunter or scavenger?

Some time between the beginning of the hominid lineage and the evolution of *Homo sapiens sapiens*, an essentially apelike behavioral adaptation was replaced by what we would recognize as human behavior, namely the hunter—gatherer way of life. How and when this occurred is of course central to paleoanthropological concerns. As we have seen, fossil evidence reveals the fundamental anatomical changes during this period, but it is to archeology that one turns for direct evidence of behavior.

The earliest stone artifacts so far recognized in the record are dated to about 2.5 million years ago, which happens to coincide closely with the earliest evidence for major brain expansion associated with the evolution of the genus *Homo*. From their earliest appearance in the record, stone tools occur both as isolated scatters and, significantly, in association with concentrations of animal bones. What exactly this association between bones and stones means in terms of early hominid behavior has in recent years become hotly debated among archeologists.

Until recently some archeologists argued by analogy with modern hunter—gatherer societies that

the associations were remains of ancient campsites, or fossil home bases, to which meat and plant food were brought to be shared and consumed amidst a complex social environment. Others have countered, suggesting that these merely indicated that hominids used the stones to scavenge for meat scraps and marrow bones at carnivores' kill sites, and therefore had no social implications whatsoever. Hence the debate, which has often been characterized as 'hunting versus scavenging' concerns how 'human' was the behavior of hominids 2 million year ago.

During the 1960s and early 1970s, paleoanthropologists considered hunting to be the primary human adaptation, a notion that has deep intellectual roots, reaching right back to Darwin's *Descent of Man*. The apogee of the 'hunting hypothesis' was marked by a Wenner—Gren Foundation conference in Chicago in 1966, titled 'Man the hunter'. The conference not only stressed the idyllic nature of the hunter—gatherer existence—'the first affluent society' as one authority had termed it—but also firmly identified the technical and organizational demands of hunting as the driving force of hominid evolution.

A shift of paradigms occurred in the mid to late 1970s, when the late Glynn Isaac proposed the 'food sharing hypothesis'. Cooperation was what made us human, argued Isaac, specifically cooperation in the sharing of meat and plant food resources that were routinely brought back to a social focus, the home base, the males having done the hunting,

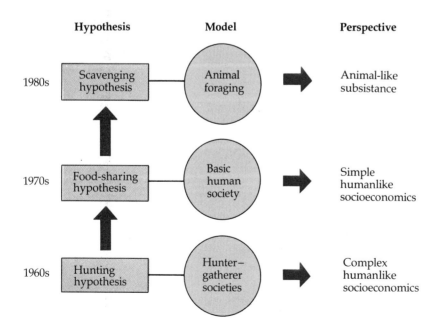

Hypothesis	Model	Perspective
1980s — Scavenging hypothesis	Animal foraging	Animal-like subsistance
1970s — Food-sharing hypothesis	Basic human society	Simple humanlike socioeconomics
1960s — Hunting hypothesis	Hunter—gatherer societies	Complex humanlike socioeconomics

An evolution of hypotheses: During the past three decades, ideas about the nature of early hominid subsistence (social and economic) activities have passed through several important stages. In the 1960s anthropologists thought of hominid evolution in terms of the impact of cooperative hunting. In the 1970s the image shifted ground somewhat, with social and economic cooperation being focused on a mixture of hunting and gathering. Current hypotheses downplay hunting as an important component in early hominid subsistence, and also throw doubt on the social complexity implied by economic cooperation. These theoretical trends have had the effect of making early hominids less human.

the females the gathering of plant foods. As for 'Man the hunter', Isaac considered that it was not possible to decide how important hunting might have been as compared with scavenging. 'For the present it seems less reasonable to assume that protohumans, armed primitively if at all, would be particularly effective hunters', he concluded in 1978.

Although the shift from the hunting hypothesis to the food sharing hypothesis changed what was perceived to be the principle evolutionary force in early hominids, it nevertheless left them recognizably human. Specifically, the conclusion that the co-occurrence of bones and stones on Plio/Pleistocene landscapes implied a hominid home base immediately invoked a hunter–gatherer social package. And, although the food sharing hypothesis was often described by proponents as merely one of many possible candidates for explaining the evolution of human behavior, it was very seductive. As Smithsonian Institution paleoanthropologist Richard Potts has observed: 'The home base/food sharing hypothesis [was] a very attractive idea because it integrates many aspects of human behavior and social life which are important to anthropologists—reciprocity systems, exchange, kinship, subsistence, division of labor, technology, and language.'

Realizing that several unspoken assumptions were built into these interpretations, Isaac in the late 1970s initiated a program of research that would test the food sharing hypothesis. Lewis Binford of the University of New Mexico independently embarked on a similar venture. Several basic issues were addressed. First, what processes brought concentrations of stone artifacts and animal bones together in particular sites? Second, if the bones and stones are causally related at these sites, what behavioral implications are possible? For Isaac and his associates, these questions were addressed by the re-examination of fossil bones from several 1.8 million-year-old sites already excavated at Olduvai Gorge, and the new excavation of a 1.5 million-year-old site at Koobi Fora, known as site 50. For Binford the exercise involved the scrutiny of published material on the Olduvai sites.

There are in fact several ways in which bone fragments and stone artifacts might accumulate at the same site and yet be causally unrelated. For instance, they might be independently washed along by a stream, and deposited together—a hydraulic jumble, as it is known. Or, carnivores

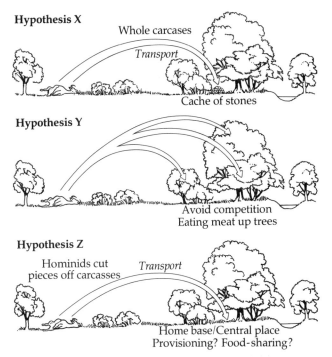

Rival hypotheses: Accumulations of stone artifacts and broken animal bones form an important element of the early archeological record. Traditionally interpreted as the remains of some kind of hominid home base (hypothesis Z), these accumulations are now subject to other interpretations. For instance, in hypothesis Y the accumulation occurs at one location because hominids used the trees there in which to escape competition from other carnivores while eating scavenged meat. Hypothesis X argues that hominids made caches of stones, to which they brought the more easily transported carcass fragments. In each case the result is the same: an accumulation of bones and stones in one location. (Courtesy of Glynn Isaac.)

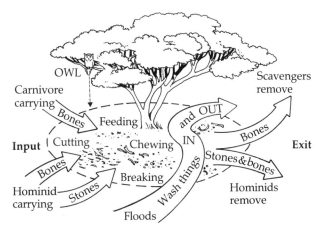

Site dynamics: Many factors influence what might be brought to a locality and what might be removed from it. Archeological excavations can recover only what remains at a site, and what can become preserved (bones and stones, not plant and soft animal material). (Courtesy of Glynn Isaac.)

might use a particular site for feeding on carcasses, while hominids might use the same site for stone knapping and whittling wood, having no interest in the bones whatsoever. The first possibility can be tested by the detailed stratigraphy of the site. The second would need some kind of indication that the stones were used on the bones in some way.

There are six major early bone and artifact sites at Olduvai Bed 1, the most famous site is the *Zinjanthropus* 'living floor', which has an accumulation of more than 40,000 bones and 2647 stones. Hydraulic processes have apparently had little or no influence in the formation of most of the sites. Binford's analysis of the sites involved comparing the pattern of bone composition with that of modern carnivore sites, the assumption being that any

difference could be attributed to hominid activity—residual analysis, it is called. His conclusion was forthright: 'The only clear picture obtained is that of a hominid scavenging the kills and death sites of other predator—scavengers for abandoned anatomical parts of low food utility, primarily for purposes of extracting bone marrow....[There] is no evidence of "carrying food home".'

For Binford, therefore, the Plio/Pleistocene bone accumulations of the oldest archeological sites at Olduvai were principally the result of carnivore activity, with hominids playing the role of a marginal scavenger. No humanlike social implications are to be made for such hominids. 'The famous Olduvai sites are not living floors', he concluded.

This last conclusion has also been reached by several of Isaac's associates, including Potts, Pat

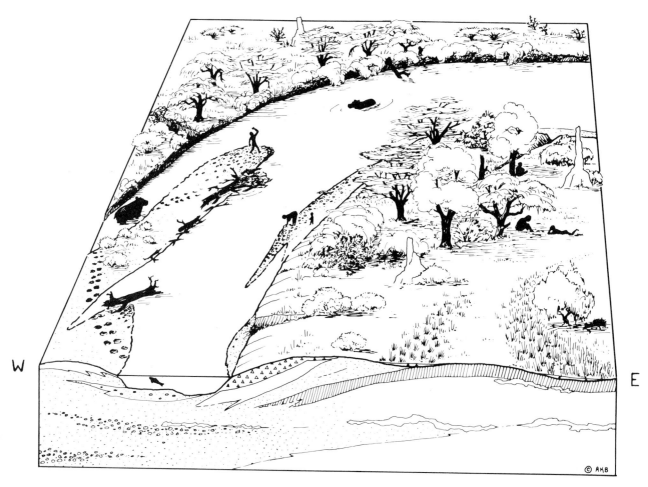

A 1.5 million-year-old site: Excavated on the floodplain east of Lake Turkana, site 50 yielded 1405 stone fragments and 2100 pieces of animal bone. One and a half million years ago the site, located in the crook of a river course, had been used over a relatively short period of time. Stone tools and the debitage struck during their manufacture could be reconstructed to form the original pebble used by the tool makers, and smashed animal bones could be conjoined to make whole bone sections. Microscopic patterns on stone tool edges indicate their use in cutting meat, soft plant material, and wood. (Courtesy of A.K. Behrensmeyer.)

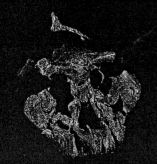

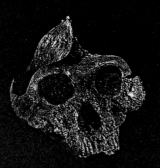

DAVID L. BRILL

E *Australopithecus aethiopicus*

Date range:
2.6 to 2.2 million
years ago.
Distribution:
Eastern Africa.
Features:
Massive chew-
ing muscles
anchored to a
prominent bony
crest along the
top of the skull.

F *Australopithecus boisei*

Date range
(including
A. aethiopicus*):
2.6 to 1 million
years ago.
Distribution:
Eastern Africa.
Features: Power-
ful upper body,
tall upper jaw,
largest molars of
any hominid.

G *Australopithecus robustus*

Date range:
2 to 1.2 million
years ago.
Distribution:
Southern Africa.
Features: A flat-
ter face, with
more prominent
cheeks and less
protruding jaws
than *afarensis* or
africanus.

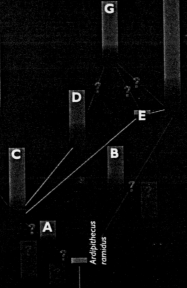

Later *Homo*
(including *H. sapiens*)

K

J

H

I

F

G

D

E

C

B

A

Ardipithecus ramidus

A *Australopithecus anamensis*
B *Australopithecus bahrelghazali*
C *Australopithecus afarensis*
D *Australopithecus africanus*
E *Australopithecus aethiopicus*
F *Australopithecus boisei*
G *Australopithecus robustus*
H *Homo habilis*
I *Homo rudolfensis*
J *Homo ergaster*
K *Homo erectus*

Hominid phylogeny according to
Bernard Wood, University of Liverpool

H *Homo habilis*

Date range
(including
H. rudolfensis*):
2.5 to 1.6 million
years ago.
Distribution:
Eastern and
southern Africa.
Features: Larger
brain and smaller
teeth than
Australopithecus.

Shipman of Johns Hopkins University, and Henry Bunn of the University of Wisconsin. However, interpretations of what exactly the bone accumulations are differ widely. Specifically, none of the three agrees with Binford that the accumulation is primarily the result of carnivore activity. All see the accumulations as the work of hominids, which carnivores occasionally visited. But Potts, Bunn, and Shipman differ among themselves as to how much of the accumulation is attributed to hunting and how much to scavenging.

Binford's analysis has been criticized on a number of grounds, not least of which, as Potts points out, is that this version of residual analysis has the *a priori* assumption that hominids displayed no carnivore-like activity: if hominids hunted and consumed animals like other carnivores do, then the resulting bone fragment pattern would be subsumed under 'carnivore activity', leaving no residual. Potts's own analysis of the Olduvai archeological sites shows that the pattern of bone accumulation is more diverse than would be expected at exclusively carnivore sites. He concludes that the accumulations are probably a mixture of scavenging and hunting, and argues that it is difficult if not impossible to distinguish between the bone accumulation patterns that would result from hunting as against what he calls early scavenging. Early scavenging could occur when, say, a hominid locates a dead animal that has not yet been partially eaten by a conventional carnivore.

In 1979 Potts, Shipman, and Bunn simultaneously discovered cutmarks on fossil bones at Olduvai, which apparently had been inflicted by stone flakes used to deflesh or disarticulate the bones. Cutmarks stand as perhaps the most direct evidence possible that hominids did indeed make use of the bones at the archeological sites. But, again, interpretations differ somewhat.

Shipman, for instance, sees little or no indication that the Olduvai hominids were disarticulating bones, and therefore concludes that the bone accumulations were principally the remains of scavenging from other carnivore kills. Both Potts and Bunn see what they interpret as evidence of disarticulation of bones, which may be taken to indicate hunting or early scavenging. Of the two, Bunn more strongly favors hunting as an important aspect of Olduvai hominids' behavior. Potts points out, incidentally, that there are very few pure hunters and pure scavengers in nature, most carnivores doing some of both. 'To ask whether early

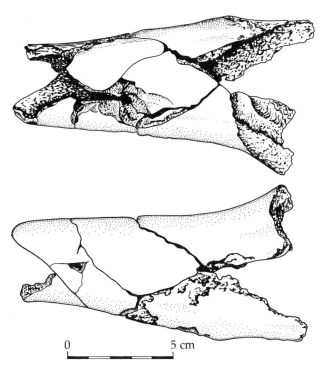

Two views of the end of a humerus of a large extinct antelope that had been shattered, presumably by occupants of Site 50: In addition to indications of precussion, this bone also bears cut marks that were probably made by sharp stone implements, such as flakes, that might have been used to remove meat or other tissues. (Courtesy of A.B. Isaac.)

hominids were hunters *or* scavengers is therefore probably not an appropriate question', he says.

Nevertheless, whether they were hunted or scavenged, the remains of animals at the Olduvai sites could in principle be an indication of hominid home bases. However, this seems unlikely. Typical hunter—gatherer home bases are places of intense social activity and havens of safety that are occupied for periods of a few weeks, and then abandoned. the Olduvai sites, however, apparently accumulated over periods of between 5 and 10 years, and were obviously visited by carnivores. The carnivores left their signatures on the sites in the form of tooth marks on some of the bones. Some of these tooth marks overlap cutmarks, which seems to imply that hominids got to the bones first. Others, however, are themselves overlapped by cutmarks, which appears to confirm that the hominids occasionally scavenged from carnivore kills.

If the Olduvai sites are not typical home bases, what were they? Potts has suggested that they formed around stone caches, places at which hominids accumulated raw material for making artifacts.

A small core/unifacial chopper: With hammerstone in position to knock off next flake (from Site 50). (Courtesy of Glynn Isaac.)

He performed computer simulations that seemed to show that, on energetic grounds, forming stone caches and bringing carcasses to them would be an optimal strategy. In any case, the raw material for the artifacts at some of the sites apparently came from sources up to 11 kilometers distant. And some of this raw material was never processed: such lumps have been called manuports.

The upshot of all this work is that the Olduvai sites appear to have been formed by hominids transporting stone to particular localities, where they probably also brought meat-bearing bones, the result of either hunting or scavenging. Instead of home bases, these sites appear to have been meat processing and consumption places. Not all early sites are the same, however. For instance, some at Koobi Fora, including site 50, are clearly not stone cache sites, because the source of the stone artifact raw material is right on the spot. Moreover, several of the stone flakes at site 50 show signs of wood whittling and processing of

soft plant material, which might imply a more leisurely use of the site than might otherwise have been envisaged. Whether this represents a change through time—site 50 being about 300,000 years younger than the Olduvai sites—or differences in ecological context is simply unknown as yet.

Isaac's response to the findings was to suggest that the food sharing hypothesis be replaced by the central place foraging hypothesis. 'Conscious motivation for "sharing" need not have been involved', he wrote in 1982. 'My guess now is that in various ways, the behavior system was less human than I originally envisaged, but that it did involve food transport and *de facto*, if not purposive, food sharing and provisioning.' CP sites, as they are sometimes known, might then be seen as being evolutionarily antecedent to true home bases.

Key questions:

● How different are the patterns of bone accumulations at the Olduvai sites from those at pure carnivore sites?
● What is the distribution of cut marks on Olduvai bones, and what does it imply about the integrity of the bones that were transported to the sites.
● Can a distinction be determined between evidence for hunting as against evidence for early scavenging?
● What kind of social organization might be implied by the central place foraging hypothesis?

Key references:

Lewis Binford, 'Human ancestors: changing views of their behavior', *Journal of Anthropological Archeology*, vol 4, pp 292–327 (1985).
Henry Bunn and Ellen Kroll, 'Systematic butchery by Plio/Pleistocene hominids at Olduvai Gorge, Tanzania', *Current Anthropology*, vol 27, pp 431–452 (1986).
Glynn Isaac, 'The archeology of human origins', in *Advances in World Archeology*, vol 3, pp 1–87 (1984).
Richard Potts, *Early hominid activities at Olduvai*, Aldine, 1988.
Pat Shipman, 'Scavenging or hunting in early hominids', *American Anthropologist*, vol 88, pp 27–43 (1986).

21 / *Homo erectus*

'It is widely accepted that populations similar to *Homo erectus* were directly ancestral to the earliest members of the living species *Homo sapiens*, although the exact timing, geography and mode of transformation are still controversial.' This statement, made recently by Eric Delson of the American Museum of Natural History, New York, pinpoints some of the key issues currently bearing on *Homo erectus*.

First discovered in 1891 by Eugene Dubois in Central Java, *Homo erectus* fossils have since been found in Africa, China, and Europe. The species appears to have evolved in Africa about 1.6 million years ago, with populations migrating first to Asia then to Europe, starting about 1 million years ago; the line became extinct something less than half a million years ago. This timing places *Homo erectus* between *Homo habilis* and the earliest appearance of *Homo sapiens*, and, as Delson notes, is widely assumed to be part of that ancestral chain. But, as we shall see later, this assumption is currently being challenged.

During the million-plus years of *Homo erectus*'s existence, a number of important 'firsts' were recorded in human prehistory: the fist appearance of hominids outside Africa; the first appearance of systematic hunting; the first appearance of anything like 'home bases'; the first systematic tool making; the first use of fire; the first important reduction of sexual dimorphism of body size among hominids; and the first indication of extended childhood. *Homo erectus* was apparently capable of a life more complex and varied than had previously been possible.

The brain size was increased over that in *Homo habilis*, ranging between 900 and 1100 cm^3. Also increased was the body size, which reached close to 1.8 meters in males and about 1.55 meters in females. The cranium itself is long and low, being somewhat flattened at the front and back, the cranial bone being thick. The jaw juts out, and the eyes are widely set, with a pronounced brow ridge above them. The postcranial skeleton, which until recently was not very well represented in the fossil record, is robust and was clearly heavily muscled.

In 1985 Richard Leakey and his colleagues reported the recovery of the remains of a remarkably complete skeleton of an approximately 12-year-old *Homo erectus* youth, which revealed some surprising anatomy. For instance, in the cervical and thoracic vertebrae, the hole through which the spinal cords runs is significantly smaller than in modern humans—presumably indicating a smaller demand for nerve signal traffic. In addition, the spines on all the vertebrae are longer and do not point as far back as in modern humans, the significance of which is puzzling.

The thigh bone is unusual, in that the femoral neck is relatively long while the femoral head—which is part of the ball-and-socket joint with the pelvis—is large. This combination is something of a mix between modern human and australopithecine anatomy: modern humans have a short femoral neck attacked to a large head, while in australopithecines the neck is long and the head is small.

The pelvis itself indicates that the birth canal was smaller than in modern humans, which implies that infants born to *Homo erectus* mothers would have needed to continue fetal growth rates after birth. This so-called secondary altricial condition means that a more extended period of child care

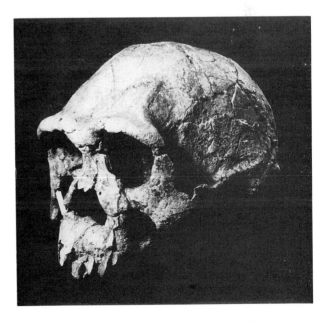

Homo erectus, **KNMER 3733:** Dated at 1.6 million years old, this specimen is the most complete and oldest of its type so far known. Note the prominent brow ridges and rounded cranium (about 850 cm^3 capacity). (Courtesy of Peter Kain and Richard Leakey.)

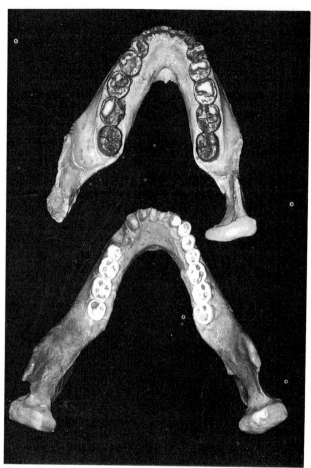

Dental trend continues: The photograph of the lower jaw of *Homo erectus* (bottom) as compared with *Australopithecus boisei* reveals the continuation of the reduction in robusticity in the chewing apparatus of *Homo*. (Courtesy of Milford Wolpoff.)

was inevitable, which might well have had important social consequences.

The *Homo erectus* youth, which came from 1.6 million-year-old deposits on the west side of Lake Turkana in Kenya, is 'the first [early fossil hominid] in which brain and body size can be measured accurately on the same individual', note Leakey and his colleagues.

One of the hallmarks of *Homo erectus* was a particular type of stone tool, the tear-drop shaped handaxe. These implements, which are usually called Acheulian handaxes after a site in France where they were first discovered, appear in 1.4 million-year-old deposits at Olduvai Gorge, contemporaneously with the appearance of *Homo erectus* specimens. Older specimens have been recovered from Koobi Fora in Kenya. Handaxes—sometimes crudely made, sometimes beautifully fashioned, and sometimes showing indications of individual or local styles—are found from these early dates right through to 200,000 years ago in Europe and Africa. Curiously, however, they were not an important feature of stone-tool technologies in East Asia.

The accumulations of bones and stones that appear in the archeological record, apparently coincidentally with the origin of the genus *Homo*, become more frequent through *Homo erectus* times and give an increasingly clear putative signal of some hunting activity (see unit 20). Some investigators speculate that a more broadly based diet, which included a greater proportion of meat than

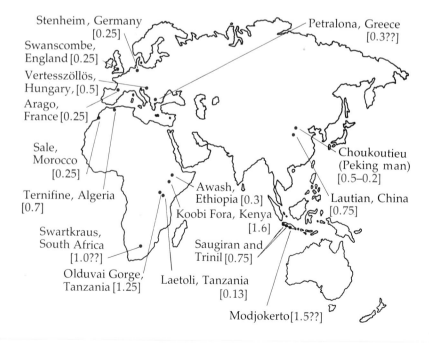

Stenheim, Germany [0.25]
Swanscombe, England [0.25]
Vertesszöllös, Hungary, [0.5]
Arago, France [0.25]
Sale, Morocco [0.25]
Ternifine, Algeria [0.7]
Swartkraus, South Africa [1.0??]
Olduvai Gorge, Tanzania [1.25]
Petralona, Greece [0.3??]
Choukoutieu (Peking man) [0.5–0.2]
Lautian, China [0.75]
Awash, Ethiopia [0.3]
Koobi Fora, Kenya [1.6]
Saugiran and Trinil [0.75]
Laetoli, Tanzania [0.13]
Modjokerto [1.5??]

Major *Homo erectus* sites: Figures in parentheses indicate age, when followed by ?? the date given is still unsettled. Note that all the sites older than 1 million years are confined to Africa, with the exception of the highly uncertain attribution of 1.5 million years to Modjokerto in Indonesia. The dispersal of *Homo erectus* from Africa to the rest of the Old World might be related to the greater territory required, and allowed, by an active predatory habit in a large primate.

was eaten by earlier hominid species, was a factor in the migration of *Homo erectus* out of Africa.

In any case, significant behavioral changes almost certainly occurred with the origin of *Homo erectus*, to judge from a comparison of body size between males and females. Sexual dimorphism in earlier hominids was large, with males being almost twice as bulky as females, a situation that has several possible behavioral implications. For instance, it might imply significant competition between males for access to females (see units 10 and 11). With *Homo erectus* this ratio dropped considerably, with males being only 20 to 30 per cent larger than females, perhaps implying a significant reduction

in competition between males. Perhaps the greater complexity of *Homo erectus* lifeways included a degree of male–male cooperation? Whether this greater complexity was predicated upon the use of a spoken language is a matter of speculation (see unit 28).

With its altogether more 'human' aspect, *Homo erectus* has long been regarded as the direct antecedent to *Homo sapiens*. Very recently, however, this assumption has been questioned. Specifically, several investigators propose that the many large-brained fossil hominids from the Middle Pleistocene that traditionally have been assigned to *Homo erectus* in fact belong to several species of *Homo*,

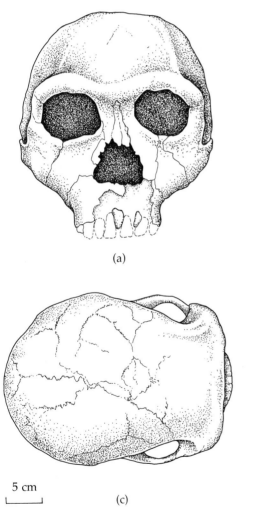

(a)

(c)

5 cm

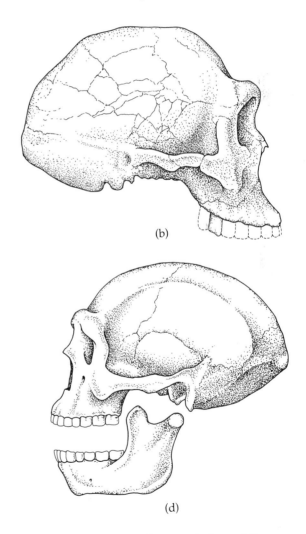

(b)

(d)

Three views (a, b, c) of *Homo erectus* skull KNMER 3733 from Koobi Fora in Kenya: The evolutionary expansion of the brain seen in *Homo habilis* is continued in this species, being about 850 cm^3 in this specimen. The pronounced brow ridges clearly distinguish the specimen as *Homo erectus*. Note that the skull comes almost to a point in the occipital (back of head) region, a feature seen in some other populations of *Homo erecuts*, particularly in Peking man (d). The Peking population lived 1 million years after those at Koobi Fora. (Courtesy of W.E. Le Gros Clark/Chicago University Press, 1955, and A. Walker and R.E.F. Leakey/*Scientific American*, 1978, all rights reserved.)

not just one. At its simplest, *Homo erectus* as currently defined from Asia would be one species, which became extinct some time in the last half million years. The second would be 'populations similar to *Homo erectus*'—Delson's words—that evolved in Africa, and were ultimately ancestral to *Homo sapiens*.

This new—and distinctly controversial—view stems from a cladistic analysis of the large collection of African and Asian fossils that traditionally have been assigned to *Homo erectus*.

'The problem of defining *Homo erectus* is that it is viewed at present as a grade of human evolution intermediate between the small brained early Pleistocene hominids and the large-brained *Homo sapiens*', notes Peter Andrews of the British Museum (Natural History), London. The term 'grade' is used to encompass a population that has reached the same adaptive stage: it does not require that the organisms belong to the same group (species). As Andrews states, 'Just because the so-called *erectus* specimens are all the same size or have similar size brains is not evidence that they belong to the same species.'

Andrews and several other investigators—including Christopher Stringer and Bernard Wood—have independently carried out cladistic analyses on the *Homo erectus* fossils, and conclude that for the most part the characters traditionally used to define the species are primitive retentions (see unit 6). 'These characters *describe* the species *Homo erectus*, because they are present on all or most known specimens,' observes Andrews, 'but they do not *define* the species, because they are also present in other hominoids.'

When the primitive characters are removed from the list of the traditional *Homo erectus* 'definition', only a small number of derived characters remains. Significantly, these characters are found exclusively within the Asian fossils, leaving the African fossils outside the group; and neither do they form a link with *Homo sapiens*. In other words, the Asian *Homo erectus* population appears to be evolutionarily separate from those hominids of a similar grade in Africa, and eventually became extinct. The African population as currently known would therefore be a so far un-named species of *Homo*, which was ancestral first to European archaic *sapiens* and later to anatomically modern humans (see unit 22).

The counter to this cladistic analysis is that anatomical differences of the sort seen among *Homo erectus* as traditionally defined are precisely what would be expected in a geographically and temporally dispersed species.

In recent years several investigators have used *Homo erectus* fossils as a test of stasis versus gradualism in hominid prehistory (see unit 4). Specifically, the measure used has been brain size: did it remain more or less stable between 1.6 and 0.5 million years ago, as argued by Philip Rightmire of the State University of New York, Binghamton? Or did it display a gradual increase through time as the species evolved towards *Homo sapiens*, as argued by Milford Wolpoff of the University of Michigan?

The debate—unresolved—is interesting, but is dogged by arguments about the appropriate fossils to use, and the certainty of dating of key fossils. And if it turns out that more than one species of *Homo* existed at this time, analyses that group Asian and African fossils would be invalid.

This latter part of human prehistory—the Middle to Late Pleistocene—is beginning to be a hot topic in paleoanthropology, including as it does the origin of modern humans (unit 22).

Key questions:

● What socioecological factors might be involved in the reduction of sexual dimorphism in body size in *Homo erectus*?
● How would one explain the robusticity of the *Homo erectus* skeleton?
● How would one explain the apparently smaller spinal cord, as seen in the west Turkana boy?
● How could the notion of a cladistic separation between Asian and African *Homo erectus* be further tested?

Key references:

Frank Brown *et al.*, 'Early *Homo erectus* skeleton from west Lake Turkana, Kenya', *Nature*, pp 788–792 (1985).
Eric Delson, 'Paleobiology and age of African *Homo erectus*', *Nature*, vol 316, pp 362–363 (1985).
G. Philip Rightmire, 'Stasis in *Homo erectus* defended', *Paleobiology*, vol 12, pp 324–325 (1986).
Milford H. Wolpoff, 'Stasis in the interpretation of evolution in *Homo erectus*', *ibid*, pp 325–328 (1986).
'The early evolution of man', *Courier Forschungsinstitut Senckenberg*, vol 69, edited by Peter Andrews and Jens Lorenz Franzen, 1984.

22 / Origin of modern humans

The origin of anatomically modern humans has long been an issue in paleoanthropology, tied up as it is with the fate of everyone's favorite caricature of cavemen, the Neanderthals. Anatomically speaking, the evolutionary shift from some kind of *Homo erectus* ancestor to *Homo sapiens sapiens* involved the decrease of skeletal and dental robusticity, modifications of certain functional—particularly locomotor—anatomy, and an increase in cranial volume. Behaviorally, the transition brought with it a more finely crafted tool technology and artistic expression (see units 25 and 26).

Broadly speaking, two opposing hypotheses have been promulgated over the decades, each of which has enjoyed majority support at different times as facts and theories have shifted anthropological opinion. One of them envisages the process as the result of widespread phyletic transformation, the other as a localized speciation event (see unit 4).

The former, termed the candelabra model by Harvard University's William Howells, proposes that ancestral *Homo erectus* populations throughout the world gradually and independently evolved first through archaic *Homo sapiens*, then to fully modern humans. This model therefore envisages multiple origins of *Homo sapiens sapiens*, and no necessary migrations. In this case, the Neanderthals are seen as European versions of archaic *sapiens*, and the model has therefore sometimes been called the Neanderthal phase hypothesis. One consequence of this widespread phyletic transformation would be that modern geographic populations would have very deep genetic roots, having separated from each other for a very long time, perhaps as much as a million years.

The second hypothesis, which Howells called the Noah's Ark model, envisages a geographically discrete origin, followed by migration throughout the rest of the Old World. So, by contrast with the candelabra model, here we have a single origin and extensive migration. Modern geographic populations would have shallow genetic roots, having derived from a speciation event in relatively recent times.

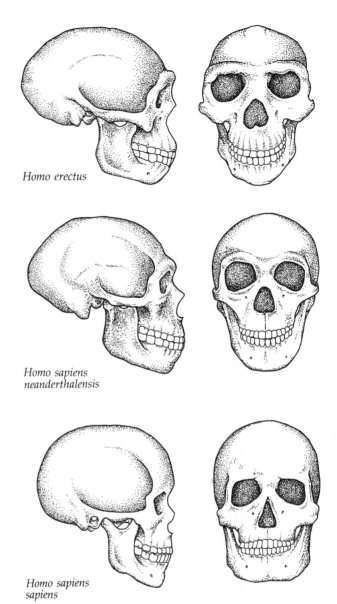

Homo erectus

Homo sapiens neanderthalensis

Homo sapiens sapiens

Comparison of *Homo sapiens neanderthalensis* with *Homo erectus* and *Homo sapiens sapiens* shows it to have some features of both: Neanderthal's robusticity, which is also apparent in the postcranial skeleton, echoes the physical appearance of *Homo erectus*, its presumed ancestor. The very large brain of the Neanderthals, which slightly exceeds that of *Homo sapiens sapiens* is seen as a modern feature. Peculiar to the Neanderthals is the extreme forward projection of the face (Courtesy of Luba Gudtz.)

If the candelabra model were correct, then it should be possible to see in modern populations echoes of anatomical features that stretch way back into prehistory: this is known as regional continuity. In addition, the appearance in the fossil record of advanced humans might be expected to occur more or less simultaneously throughout the Old World. By contrast, the Noah's Ark model predicts little regional continuity and the appearance of modern humans in one locality before they spread into others.

As presented here, these two models represent extremes, of course, and it is possible to envisage intermediates. For instance, there might have been a single geographic origin as predicted by the Noah's Ark model, but followed by migrations in which newcomers interbred with locally established populations of archaic *sapiens*. And there could have been more extensive gene flow between different geographic populations than is allowed for in the strict candelabra model, producing closer genetic continuity between populations (this is known as the 'center and edge' hypothesis). In both cases the result would be a much less clear-cut signal in the fossil record.

Until relatively recently there was a strong sentiment among anthropologists in favor of extensive regional continuity. In addition, western Europe tended to dominate discussions, for several cogent reasons. First, the best fossil record for the immediate pre-modern era comes from this part of the world: namely the remains of more than 300 Neanderthal individuals found in sites from western

Europe to the Near East. Second, there are many splendid archeological sites that, in combination, appear to document the transition of pre-modern, Middle Paleolithic, tool technologies to the more extensive and sophisticated Upper Paleolithic technologies of modern humans, about 32,000 years ago. Last, the spectacular painted caves of southern France and northern Spain dazzled investigators into believing that Europe was where all the action was in terms of the origins of *Homo sapiens sapiens*.

Evidence pertinent to the origin of anatomically modern humans has expanded considerably in recent years, and now includes molecular biology data as well as fossils. Although there is still a spectrum of interpretations of the fossil evidence, there is a distinct shift in favor of some version of the Noah's Ark model. And the molecular biology data—specifically, mapping and sequencing of nuclear genes and mitochondrial DNA—strongly support this model. (See unit 23 for a discussion of the mitochondrial DNA work.) Instead of being in the region of 35,000 to 45,000 years ago, the origin of *Homo sapiens sapiens* is now widely thought to be in excess of 100,000 years ago, according to both fossil and molecular biological evidence.

Candelabra

No migration, no replacement

Europe   Africa   Asia

Present

0.5 m

1.0 m

Regional continuity of anatomy

Noah's Ark

Migration, replacement

Europe   Africa   Asia

No regional continuity of anatomy

Modern *sapiens*

Archaic *sapiens*

Homo erectus

Two views of modern human origins: These diagrams represent the extremes of the debate over the origin of modern humans. The candelabra model (left) sees modern human populations throughout the globe as the direct, geographical descendants of populations of *Homo erectus* that moved out of Africa 1 million years ago: Asian *Homo erectus* became Asian *Homo sapiens*, European *Homo erectus* became European *Homo sapiens*, and so on. The candelabra model is characterized by regional continuity of anatomical characteristics. In the Noah's Ark model *Homo erectus* migrations from Africa 1 million years ago established populations throughout the Old World, but these were replaced by recent migrations of anatomically modern humans, also migrating from Africa. In this model there is no regional continuity. Most anthropologists support models somewhat between these two extremes.

Comparison of the Neanderthal skull with a modern human skull: The triangle in the Neanderthal skull (left) shows the spatial relationships between the forward edge of the first molar (C), the lower edge of the cheekbone (A), and the upper edge of the cheekbone. A similar relationship drawn in a modern human skull (right, with a Neanderthal outline shaded in) produces a much more flat triangle, thus illustrating the significant forward protrusion in the Neanderthal face.

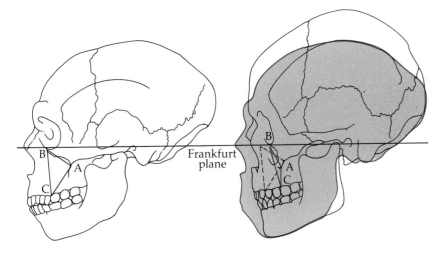

Neanderthal roots can be traced back more than 250,000 years, with the 'classic' Neanderthal anatomy becoming firmly established about 125,000 years ago, and being geographically restricted to Eurasia, specifically the region between western Europe and the Near East. In common with *Homo erectus* and other archaic *sapiens* forms, Neanderthals were skeletally very robust and heavily muscled, but were characterized by an extremely protruding mid-facial region—as if someone had taken a face made of rubber, and pulled on the nose. The brain size of Neanderthals was equal to or just slightly greater than the modern average.

Several explanations have been advanced for the Neanderthal mid-facial architecture. These include: an adaptation for warming inhaled frigid air as it passed through the enlarged nasal cavities; a means of condensing, and therefore conserving, moisture in exhaled breath; and a secondary consequence in the facial region of severe chewing pressures centered at the front of the jaw.

The Neanderthal pelvis also seems to be highly characteristic, so much so that in incomplete specimens the pelvic canal appeared to be unusually large, prompting University of New Mexico anthropologist Erik Trinkaus to postulate that gestation was prolonged in this species, the infant at birth therefore being larger than in modern humans. However, when a more complete specimen came to light in 1987—from Kebara, in Israel—it was found that the pelvic canal is not unusually large, merely that the pubic bone is extraordinarily long.

This, and several other features not seen in modern human pelvices, supports the increasingly popular conclusion that Neanderthals represent a dead end in human evolution, not a stage on the way to *Homo sapiens sapiens*. On this basis there is

argument for removing the Neanderthals from the subspecies status of *Homo sapiens neanderthalensis* and instead reverting to the *Homo neanderthalensis* nomenclature.

According to last appearances in the fossil record, Neanderthals disappeared in a wave flowing east to west, between 45,000 and 32,000 years ago. Most authorities now believe this disappearance to be extinction as a result of replacement—at least in the west—by incoming anatomically modern humans, whose overall anatomy is characteristic of equatorial adaptation, not cold adaptation. Complete replacement of one population by another need not be as dramatic a process as might be envisaged, as Ezra Zubrow of the State University of New York at Buffalo has calculated. A subsistence advantage of just 1 per cent by modern humans in competition with Neanderthals, for instance, could result in complete replacement within 30 generations, or a millennium. Nevertheless, some investigators argue that interbreeding took place, infusing modern populations with some Neanderthal characters.

In addition to the classic Neanderthals, the fossil evidence most immediately relevant to the origin of modern humans is to be found throughout Europe, Asia, Australasia, and Africa, and goes back in time as far as 300,000 years. Most of these fossils—which principally are crania of varying degrees of incompleteness—look like a mosaic of *Homo erectus* and *Homo sapiens sapiens*, and are generally termed archaic *sapiens*. It is among such fossils that signs of regional continuity are sought, being traced through to modern populations.

For instance, Milford Wolpoff of the University of Michigan argues for such regional anatomical continuities among Australasian populations and

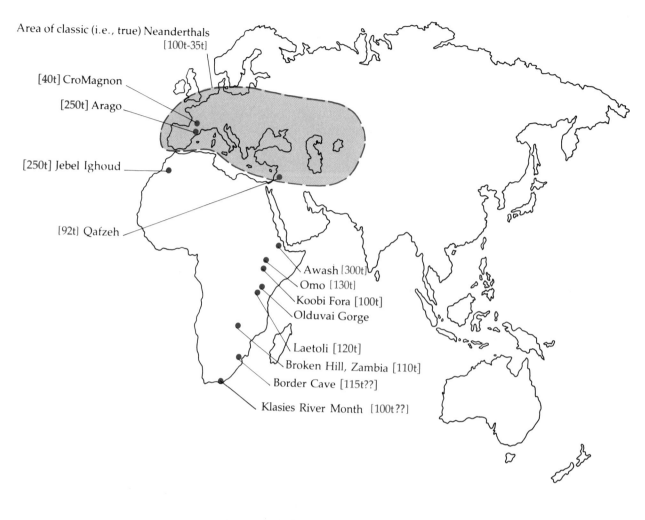

Area of classic (i.e., true) Neanderthals [100t-35t]

[40t] CroMagnon

[250t] Arago

[250t] Jebel Ighoud

[92t] Qafzeh

Awash [300t]
Omo [130t]
Koobi Fora [100t]
Olduvai Gorge

Laetoli [120t]
Broken Hill, Zambia [110t]
Border Cave [115t??]
Klasies River Month [100t??]

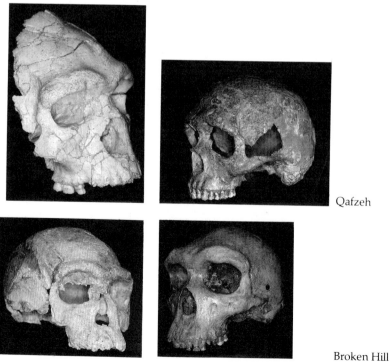

Arago face

Qafzeh

Jebel Irhoud

Broken Hill

among Chinese populations. Citing different characters, Fred Smith of the University of Tennessee believes a good case can be made for regional continuity in Central Europe and perhaps in North Africa. Proponents of a replacement model argue that, in most of these cases, the anatomical characters being cited as indicating regional continuity are primitive (see unit 6), and therefore cannot be used uniquely to link specific geographic populations through time. Peter Andrews and Christopher Stringer of the British Museum (Natural History), London, are prominent among those scholars making this argument.

If the archaic *sapiens* populations were not part of a general phyletic transformation towards *Homo sapiens sapiens*, what were they? It is possible that they represent several different species, as Ian Tattersal of the American Museum of Natural History, New York, has argued. This is not a popular notion, but it is undeniable that this group of fossils is something of a rag-bag that begs to be sorted out.

The equatorial anatomy of the first modern humans in Europe presumably is a clue to their origin: Africa? There are sites from the north, east and south of the continent with varyingly questioned claims on anatomical modernity and geological antiquity. Perhaps the one most accepted on both counts is Klasies River Mouth, South Africa. 'The fossils from this site are totally modern in all observable respects,' comments Richard Klein of the University of Chicago, 'including the presence of a strongly developed chin'. The dating, however, remains to be nailed down, though a range of 80,000 to 115,000 years is strongly supported.

Does this mean that *Homo sapiens sapiens* arose by a speciation event in southern Africa, populations migrating north, eventually to enter Eurasia? This is a clear possibility, but a key fossil in such a scenario comes from a cave site on the southwest flank of Mount Qafzeh in Israel. Undoubtedly

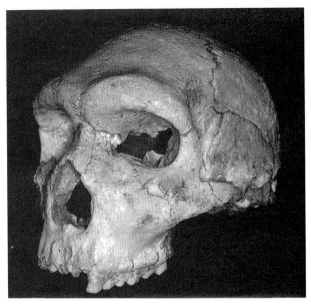

The Petralona skull: Recovered from a cave in Greece and dated between 300,000 and 400,000 years old. Although the face is partially eroded, the cranium can be seen to be a mosaic of *Homo erectus* and *Homo sapiens* features. For example, the brow ridges are ancient, while the large rounded cranium is evidence of modern development.

anatomically modern, the age of the Qafzeh cranium has been in some doubt for years, but was recently reported to be 90,000 years old. If correct, such a date implies at the very least that anatomically modern humans and Neanderthals overlapped—if not directly coexisted—in this part of the world for at least 45,000 years. In addition, it might imply that modern humans arose in North Africa, and spread both north and south. Only when more fossils in this part of the world are firmly dated will this issue be settled.

Overall, therefore, a strong case can be made for the replacement model in western Europe, while in the rest of the Old World the issue remains to be resolved. Nevertheless, the earlier appearance of anatomically modern humans in Africa than in Europe and probably Asia too supports the Noah's Ark hypothesis. The degree of interbreeding that might have occurred between invading populations of anatomically modern humans and existing populations of archaic *sapiens* is addressed by mitochondrial DNA data (unit 23).

Molecular biological evidence from mapping and sequencing nuclear DNA yields two main conclusions: the first addresses the depth of our genetic routes, the second the possible location of the origin of anatomically modern humans.

Modern human origins (facing page): The map shows the geographical distribution (and age in thousands of years) of some of the fossils that are important in pinpointing the origin of anatomically modern humans. Sites considered to be key in the equation are Border Cave and Klasies River Mouth Cave, in South Africa, and Qafzeh, Israel. With specimens that are undoubtedly modern in form, these sites are dated at 115,000, about 100,000, and 92,000 respectively (and with increasing certainty). These sites imply an origin of modern humans in excess of 100,000 years ago. The shaded area represents the distribution of classics Neanderthals.

The amount of genetic variation throughout all modern human populations is surprisingly small, and implies a recent origin for the common ancestor of us all. The argument that genetic variation among widely separated populations has been homogenized by gene flow (interbreeding) is simply not tenable, argues population geneticist Shahani Rouhani, of University College, London. For instance, he calculates that it would require almost half a million years for an advantageous gene to travel from South Africa to China by the normal process of gene flow. Rouhani points out that, in general, population genetics theory does not support the notion of phyletic transformation in populations as widespread as those of hominids in the Middle Pleistocene.

Although genetic variation is small overall, it is greatest in African populations, implying they are the longest established. Moreover, data on the β-globin cluster implies that the anatomically modern humans who migrated out of Africa went through a population bottleneck, according to James Wainscott and his colleagues at the University of Oxford. The data from Wainscott's laboratory have been joined by those from several other laboratories in supporting the Noah's Ark model, with Africa as the source.

One consequence of placing the origin of *Homo sapiens sapiens* at 100,000 years or earlier is an apparent uncoupling of modern anatomy from modern behavior. Modern tool technology—Upper Paleolithic in Europe, Late Stone Age in Africa—occurs between 35,000 and 50,000 years ago, long after the origin of the species responsible for it. It remains to be discovered what this apparent uncoupling means in terms of human prehistory. But in practical terms, it seems that, unless the picture changes dramatically, archeological evidence can no longer be adduced as a signal for the origin of modern humans.

Key questions:

● How does the current debate over the origin of modern humans fit into past ideas about the shape of the human family tree?
● What kind of socioecological data might have an impact on a choice between the candelabra and Noah's Ark models?
● What kind of evidence is mostly likely to settle the question of gene flow among Late Pleistocene populations?
● What might underlie the apparent uncoupling between the origin of anatomically modern humans and the advent of modern tool technologies?

Key references:

J.S. Jones and S. Rouhani, 'How small was the bottleneck?' *Nature*, vol 319, pp 449–450 (1986).

P. Mellars and C.B. Stringer (editors), *The origin and dispersal of modern humans*, Edinburgh University Press, 1988.

Fred H. Smith, 'Continuity and change in the origin of modern *Homo sapiens*', *Zeitschrift fur Morphologie und Anthropologie*, vol 75, pp 197–222 (1985).

C.B. Stringer and P. Andrews, 'The origin of modern humans', *Science*, vol 239, pp 1263–1268 (1988), and reply by M.H. Wolpoff *et al.*, *ibid.*, vol 241, pp 772–773 (1988).

Ian Tattersall, 'Species recognition in human paleontology', *Journal of Human Evolution*, vol 15, pp 165–175 (1986).

23 / Mitochondrial Eve

In addition to the fossil evidence discussed in the previous section, molecular biological data can also address the question of the location, timing, and subsequent pattern of dispersal of anatomically modern humans. Several research groups have for some time been examining protein and, more recently, nuclear DNA data from geographically disparate populations, looking for patterns of relatedness that might indicate an original source of *Homo sapiens sapiens*. Although initial interpretations were often weak and sometimes contradictory, a consensus has recently been building that supports the notion of an African origin. The most recent contribution to this body of data is from mitochondrial DNA comparisons, the results of which have led to declarations such as: Mitochondrial Eve, mother of us all, lived in Africa 200,000 years ago.

Although there has been a good deal of confusion surrounding the work, the data to be garnered from mitochondrial DNA are undoubtedly a potentially powerful tool for tracing human lineages through the relatively recent period of time during which *Homo sapiens sapiens* is believed to have evolved. In essence, mitochondrial DNA is a fast-ticking molecular clock. There are, however, many unresolved questions about the basic data from the mitochondrial DNA itself and about the interpretation of those data in relation to the possible population dynamic patterns that might have accompanied the origin of modern humans.

The most parsimonious interpretation of the data as they were initially presented by Allan Wilson and his University of California, Berkeley, colleagues during 1987 was as follows. *Homo sapiens sapiens* evolved from a population of archaic sapiens about 150,000 years ago somewhere in Africa. By 100,000 years ago groups of modern humans had begun to move out of Africa and establish populations throughout the Old World. There was very little, if any, interbreeding between advancing modern human populations and existing archaic sapiens populations in any part of the Old World: the roots of modern geographic populations are therefore very shallow indeed. In other words, the mitochondrial DNA data as presented here support the Noah's Ark, total replacement hypothesis described in the previous section.

These conclusions are by no means universally accepted and have been criticized on two main grounds. The first is on the reliability of the mitochondrial clock, specifically that it does not tick with a steady, average rate. The second is the suggestion that Wilson and his colleagues have incorrectly calculated the overall rate of the clock, giving an erroneously fast figure. A slower rate would yield deeper genetic roots for modern geographic populations.

Mitochondrial DNA is potentially valuable for reconstructing recent genealogies, for two principal reasons. First, mitochondrial DNA accumulates mutations about 5 to 10 times faster than does

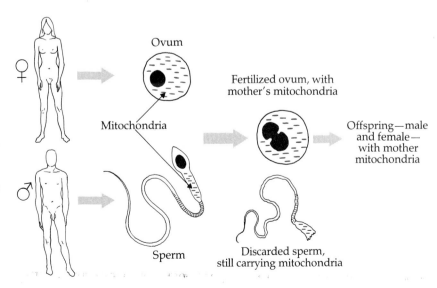

Ovum

Fertilized ovum, with mother's mitochondria

Mitochondria

Offspring—male and female— with mother mitochondria

Sperm

Discarded sperm, still carrying mitochondria

Pattern of inheritance: Unlike nuclear DNA, which we inherit half from our mother and half from our father, mitochondrial DNA is passed on only by females. The reason is that when the sperm fertilizes the egg, it leaves behind all its mitochondria: the developing fetus therefore inherits mitochondria only from the mother's egg.

nuclear DNA: it therefore effectively magnifies the passage of time. Second, in the fertilization of an ovum at conception, male sperm contribute no mitochondria to the next generation. Mitochondrial DNA is therefore inherited only through the maternal line, with no mixing of genes through recombination, thus giving a potentially clear link back through the generations.

Since Wilson's Berkeley laboratory first began applying the mitochondrial DNA technique to human origins in the late 1970s, several research groups have independently tackled the problem. Nevertheless, the largest dataset published to date is from Berkeley, which Wilson, Rebecca Cann, and Mark Stoneking produced in 1987.

The aim of the technique is twofold. First, to

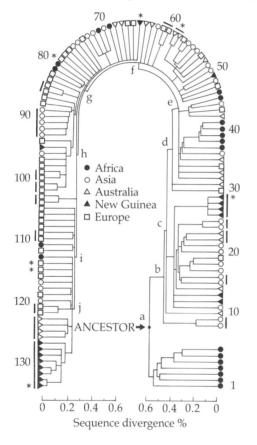

Patterns of relatedness: This 'christmas tree' shows the genetic divergence among 134 individuals from different geographic populations, whose mitochondrial DNA was tested. The tree shows a split between African and non-African populations; and the African population is the longest established, indicating the origin of modern humans in that continent. The different degrees of sequence divergence among the nonAfrican populations gives some indication on when different parts of the Old World were colonized. (Courtesy of Rebecca L. Cann *et al./Nature*.)

obtain some idea of the degree to which mitochondrial sequences vary among individuals, specifically among individuals from different geographic populations: the degree of variation in principle measures the length of time populations have been separated. Second, to reconstruct a genealogical tree based on the patterns of variation obtained.

Complete sequences of the mitochondrial genomes would provide the ultimate comparison of variability. However, even though the mitochondrial genome of humans is of modest size compared with most chromosomes—comprising 16,569 basepairs—sequencing sufficient numbers of them for a multipopulation comparison would be a gigantic task with current sequencing technology. Instead, what Wilson and other researchers do is produce a kind of map of the genome, which effectively samples about 9 per cent of its sequences.

The DNA is cut with a suite of 12 so-called restriction enzymes, each of which cleaves the DNA chain at discrete combinations of nucleotides on the chain. If everyone's mitochondrial genomes were identical in sequence, the fragments produced by this procedure would give the same pattern *in all cases*. Any variation in DNA sequence that happens to alter one of the restriction enzyme cut sites, however, will result in a different fragment pattern. It is these differences that researchers use to get a measure of sequence variability between individuals and populations.

Wilson, Cann, and Stoneking's 1987 paper presented data on 147 individuals representing five geographic localities: Africa (mainly black Americans), Asia, Europe, Australia, and New Guinea. The 147 individual genomes generated 133 different fragment patterns within which the overall degree of sequence variation was relatively small (0.57 per cent). This small degree of overall variation, compared to that in, for instance, great apes, is an immediate clue to potentially shallow genetic roots for modern human populations.

The 133 different types can be organized into a genealogical tree by parsimony analysis (looking for the smallest number of changes that will link the different patterns and eventually trace back to a common ancestor). In fact, there were many minimal or near-minimal length trees—a result of the small gradations of differences between the individual types—but all produced two primary branches. One of the branches contained mitochondrial DNA patterns from Africans only, while the other carried all the other types, plus some

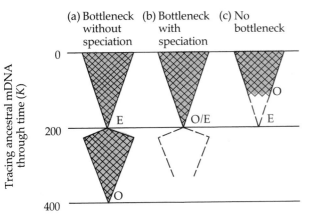

Tracing ancestral mDNA through time (K)

(a) Bottleneck without speciation (b) Bottleneck with speciation (c) No bottleneck

Population models: There are at least three population scenarios that might produce a pattern of mitochondrial DNA data that point to 200,000 years for the time of Eve (E). In (a), the origin (O) of modern humans is seen at 400,000 years, but the population goes through a severe bottleneck at 200,000 years ago. In (b) the population bottleneck gives rise to the origin of modern humans. And in (c), there is no bottleneck at 200,000 years, and the origin of modern humans is later.

African types too. The simplest interpretation of such a pattern is that the common ancestor of all mitochondrial DNA types was African.

A second inference from the tree is that each geographic region outside Africa appears to have been colonized by more than one migration. This is indicated by the fact that within each region the different types can be grouped into 'clans' in which the mitochondrial DNA is very similar. For instance, Europe has at least 36 different mitochondrial DNA clans, which might equate with as many groups of colonists. For Asia the number is 31; 14 for New Guinea; and 15 for Australia.

Differences in the degree of DNA variation among the various geographic groups also points to an African origin. Specifically, variability among the African types is significantly greater than types from other regions. If variation accumulates steadily and equally among all geographic types, then the greater degree of variability within the African group indicates that it is the longest established. In other words, DNA variability joins with genealogical analysis in fingering Africa as the source of all modern human mitochondrial genomes.

The overall degree of variability addresses the question of interbreeding versus replacement, as modern humans spread throughout the Old World. These existing populations of archaic sapiens would have been established for as long as a million years: their mitochondrial DNA would therefore be widely

different from that of the African population that gave rise to *Homo sapiens sapiens*. If modern humans did indeed interbreed with these established populations, then there would have been a mixing of mitochondria too. In which case some mitochondrial DNAs in modern populations would be very different from others. So far this has not been detected, implying that mixing did not take place.

Much of this work is of course predicated on the assumption that mitochondrial DNA accumulates mutations at a steady rate, that the clock ticks steadily. Wilson argues that the rate of mitochondrial DNA mutation is indeed steady on average, principally because most mutations affect nucleotides in noncoding regions, and are therefore probably neutral. Also the DNA data themselves demonstrate regularity, he says. Nevertheless, it is known that mutations other than neutral base changes do occur, and at very different rates. How steady the mitochondrial DNA clock is overall will be settled only when a sufficient number of complete genome sequences have been compared.

Wilson and his colleagues calibrate the rate of change by looking at the amount of change that has accumulated within clans of New Guinea people since they arrived there some 40,000 years ago. The figure they come up with is really a range: between 2 and 4 per cent, or between 2 and 4 base changes per 100 bases, per million years. These figures have been questioned, particularly by Masatoshi Nei of the University of Texas. Nei believes a 1 to 2 per cent range is more likely.

Using their own rate figures, and the 0.57 per cent mitochondrial DNA diversity among modern populations, Wilson and his colleagues calculate that the common ancestor of modern mitochondrial DNA populations lived some 140,000 to 280,000 years ago, in Africa. The figure of 200,000 years is usually given as an average.

What does this figure actually mean in terms of the origin of *Homo sapiens sapiens*? If, as some people favor, there was a population bottleneck associated with the this speciation event—possibly crashing to as little as a single mating pair—then the Mitochondrial Eve of 200,000 years ago would also represent the very beginning of the modern human lineage.

A population bottleneck, however, is not the only means by which all mitochondrial DNA genomes in modern populations might have derived from that of a single female living 200,000 years ago. It is quite possible that this single female was a member of a large population, and her mitochondria were the only ones that survive in modern populations. Based on the work of John Avise of the University of Georgia, the scenario by which this could come about is best thought of by analogy.

Imagine a population of, say, 10,000 mating pairs, each of which has a different family name, and each of which on average contributes two children to the next generation. Because the family name is inherited only through one sex (the male), a proportion of the names will be lost each generation because some families will have no male offspring. Avise calculated that after about 10,000 generations all but one of the names would have been lost. The same principle applies to mitochondrial DNA lines, says Wilson. In this case, the evolution of *Homo sapiens sapiens* might have occurred in a large population—a scenario that many population geneticists favor—some time after Eve had lived. In this case, Eve would have been an archaic *sapiens*, not *Homo sapiens sapiens*. This is the system that Wilson favors, which is why he gives a date of 150,000 years as the origin of *Homo sapiens sapiens*,

with Eve having lived at 200,000 years: this interpretation fits most closely with the fossil evidence as it is now known.

There is a third scenario—and probably others too—in which Eve and the origin of *Homo sapiens sapiens* do not coincide. If, for example, speciation had occurred at 400,000 years ago, and the population suffered a severe bottleneck at 200,000 years ago, then the pattern of mitochondrial DNA would look just as it does now. Distinguishing between these various possibilities on the genetic evidence alone would depend upon looking for signs of population bottlenecks in mitochondrial DNA variability and comparing it with signs in nuclear DNA, which would be affected differently. So far this has not been done.

Although some researchers are comforted by the apparent coincidence between the molecular and fossil data both indicating an African origin of modern humans in excess of 100,000 years, it would surely be more satisfactory if both sets of evidence were to stand alone. That way they would provide a test for each other rather than—perhaps illusory—support.

Out of Africa: According to interpretations of the mitochondrial DNA data, modern humans arose somewhere in Africa about 150,000 years ago. About 100,000 years ago populations moved into the rest of the Old World, reaching Australasia about 50,000 years ago, and western Europe 35,000 years ago. Asia appears to have been settled by 31 'clans', Europe by 36, and Australia by 15.

Key questions:

- Does the mitochondrial DNA clock tick regularly?
- What would be the implication of a slower rate of divergence of mitochondrial DNA in two lineages than the 2 to 4 per cent that has been calculated?
- What is the likely relationship between Eve and the origin of anatomically modern humans?
- Do the mitochondrial DNA data exclude the possibility of interbreeding between *Homo sapiens sapiens* and existing archaic *sapiens* populations?

Key references:

R. Cann *et al.*, 'Mitochondrial DNA and human evolution', *Nature*, vol 325, p 31 (1987).
J. S. Jones and S. Rouhani, 'How small was the bottleneck?', *Nature*, vol p 319, 449 (1986).
Masatoshi Nei, 'Human evolution at the molecular level', in *Population Genetics and molecular evolution*, edited by T. Ohta and K. Aoki, Springer-Verlag, 1985, p 41.

24 / Tool technology: the early cultures

The use of tools had a tremendous impact on the path of human prehistory, specifically in allowing access to food not otherwise available, and their beneficial economic rewards played an important part in the ultimate biological success of the ancient tool makers. Because stone does not perish through archaeological time, whereas hide, tendons, wood and bark rarely become fossilized, the prehistoric record is heavily biased toward stone-tool technology, in the early stages at least. Digging sticks and wooden spears may well have played an important, perhaps even dominant, role in early technologies, but the record is virtually silent on the matter.

Recently, however, the work by Lawrence Keeley and Nicholas Toth on wear on stone tools from Kenya provided a useful reminder of the probably common use of wooden implements in early times. According to their analysis, one of a series of small stone flakes from a 1.5 million-year-old site shows distinct signs of woodworking, as if it had been used in whittling a stick. Whether the finished product was a spear, a digging stick or some other implement, one will never know. But the discovery is important, not least because it shows the use of one tool in the preparation of another.

The earliest putative stone artifacts discovered so far come from Ethiopia and are dated at around 2.5 million years. They are a collection of extremely crude scrapers, choppers, and flakes, each the product of a very few blows with a hammerstone. In archeological terms, they are described as an example of the Oldowan industry. Looking forward through time from this earliest example of tool making, one gains two powerful impressions.

First, there is a striking continuity through vast tracks of time. Tools such as these earliest artifacts represent the dominant form of stone-tool technology for more than a million years. About 1.5 million years ago a new industry emerges, which is known as the Acheulian. This industry represents only a modest advance over the Oldowan, and is characterized by the presence of tear-drop shaped handaxes. The Acheulian did not replace the Oldowan immediately, but merely accompanied it through half a million years of human history, after which it became the dominant form. Even so, tools that can be described as Oldowan in type were still to be found in eastern Asia right up to 200,000 years ago and less. In Africa and Europe the Acheulian continued as the main tool industry, until it too began to be replaced around 150,000 years ago.

The second powerful impression of stone-tool

Representative examples of Oldowan tools: Top row: hammerstone; unifacial chopper; bifacial chopper; polyhedron; core scraper; bifacial discoid. Bottom row: flake scraper; six flakes. An actual tool kit on site would mainly comprise flakes. (Courtesy of Nicholas Toth.)

Representative examples of Acheulian tools: Top row: ovate handaxe; pointed handaxe; cleaver; pick. Bottom row: spheroid (quartz); flake scraper; biface trimming flake; biface trimming flake. (All artifacts, except spheroid, are lava replicas made by Nicholas Toth.) (Courtesy of Nicholas Toth.)

113

technologies up to about 150,000 years ago is their essentially opportunistic nature. Although there is a gradual imposition of form and style through that great swath of time, it is really rather minimal. Only after 150,000 years ago is there a strong sense of stylistic order.

The Oldowan technology has been classified into a number of different putative stone-tool types, largely through the work of Mary Leakey, who excavated for several decades at Olduvai Gorge, after which the technology is named. Some half a dozen implements have been named, including discoids, spheroids, polyhedrons, core scrapers, flake scrapers, hammerstones, and pebble choppers, the last of which predominate. According to Toth, however, although such artifacts may have had specific uses, they are just as likely to be the by-product of the principal tool of the Oldowan technology: namely, the flake. Toth has shown that, depending on the shape of the starting material—usually a lava cobble of some sort—any one of the Oldowan tools can be produced automatically, simply by the systematic striking of flakes.

The oldest levels at Olduvai date back to almost 2 million years ago, and it is here that the Oldowan industry begins. The industry continues for 1 million years and more, but becomes a little more refined, adding a few more tool categories, such as awls and protobifaces. These advances, which appear about 1.5 million years ago, are recognized as the Developed Oldowan. At about the same time, a new industry, the Acheulian, appears in the record. The handaxe, as has been noted, is the hallmark of this new industry, and it represents the first tool in which a predetermined shape has been imposed on a piece of raw material.

The principal invention of early stone-tool technologies was that of concoidal fracture: strike a core at an angle and a flake, large or small, is removed. The resultant tool is very much determined by the shape of the starting material. However, the bifacial symmetry of the handaxe, with its two sharp converging edges, required the shape to be 'seen' within the lump of stone, which is then worked towards with a series of careful striking actions.

Some handaxes of later times were esthetically pleasing products of hours of skilled labour. Exactly what they were used for is still something of a mystery, but the combination of a long sharp edge with bulk and weight makes them exceedingly efficient at slicing through even the toughest hide,

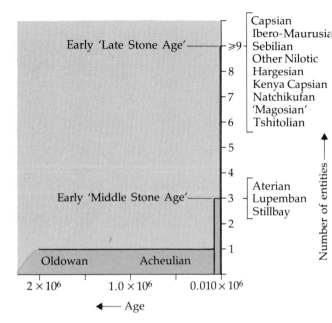

Increase in styles of tool technologies: Compared with later industries in Africa—the Middle Stone Age and Later Stone Age—the long eras of the Oldowan and Acheulian technologies each formed a single entity; there was no clear diversification into different styles. (Courtesy of Glynn Isaac.)

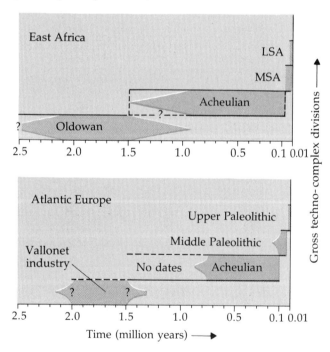

Time duration of tool technologies: With the passage of time each successive tool industry is of ever shorter duration. The element of stability in the early industries—first the Oldowan then the Acheulian—is striking. Later industries—the Middle Stone Age and Later Stone Age (MSA and LSA) in Africa and the Middle and Upper Paleolithic in Europe—pass fleetingly by comparison. (Courtesy of Glynn Isaac.)

The full set of artifacts from a small site on the eastern shore of Lake Turkana: The collection shows that the majority of pieces, as in most sites, are small, sharp flakes. (Courtesy of Glynn Isaac.)

including that of elephants and rhinoceroses. Cleavers, which were also part of the Acheulian technology, had similar properties, but are cruder, less elegant implements.

The Acheulian industry, which had about 10 principal implements, continued for more than a million years, before it was replaced by the wide range of much more refined tools of the Mousterian culture: the products of *Homo sapiens neanderthalensis*. During that long period of time the best examples of the Acheulian technology became increasingly more elegant, but throughout this time there were crude examples, similar to those from the beginning of the record.

Glynn Isaac points out that from the very beginning of stone-tool making the range of implements produced does not increase significantly. What does change through time, however, is the degree of standardization, the frequency of producing certain forms against a background of 'noise'. *Ad hoc* stone

knapping gives way to deliberate imposition of pre-conceived order.

The duration of the Acheulian saw certain idiosyncratic expression, for example in details of the shape of tools and their size. Differences in availability of suitable stone, different specific technological needs and an element of individual style would have contributed to this. There was, however, a certain homogeneity at any particular time. There was, in a sense, just one Acheulian culture. From 150,000 years onwards, this pattern of culture began to change, at an ever accelerating pace. It is, as Isaac says, as if some threshold was passed: 'a critical threshold in information capacity and precision of expression.'

Key questions:

● What kinds of preconceptions might influence the interpretation of the earliest stone-tool technologies?
● Compared with tool use in apes, how important an innovation is the production of a stone flake?
● What are the implications of the stasis and opportunistic nature of early stone-tool technologies?
● How would one test the notion that only members of the genus *Homo* made and used stone tools?

Key references:

Glynn Isaac, 'The archeology of human origins', *Advances in world archeology*, vol 3, pp 1–87 (1984).
Lawrence H. Keeley, 'The functions of Paleolithic tools', *Scientific American*, November 1977.
richard Klein, 'Stone age prehistory of Southern Africa', *Annual Review of Anthropology*, vol 12, pp 25–48 (1983).
Nicholas Toth, 'The first technology', *Scientific American*, April 1987, pp 112–121.

25 / Tool technology: the pace changes

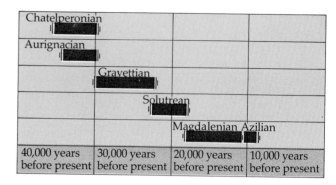

Chatelperonian ▮▮			
Aurignacian ▮▮			
	Gravettian ▮▮▮		
		Solutrean ▮▮	
		Magdalenian ▮▮▮	Azilian ▮
40,000 years before present	30,000 years before present	20,000 years before present	10,000 years before present

Tool industries of the Upper Paleolithic: The pace of change of tool technologies becomes almost hectic from 40,000 years onwards. In addition, the tool industries themselves take on a complexity and refinement unmatched in earlier periods. A distinct sense of fashion and geographic variation is also well developed.

Solutrean laurel leaf blade: Some examples of which are so thin as to be translucent. They were probably used in rituals rather than in practical affairs. (Scale bar: 5 cm.) (Courtesy of Roger Lewin and Bruce Bradley.)

Between 250,000 and 150,000 years ago the pace of change of tool technologies began to accelerate. Whereas continuity was the hallmark of tool making prior to this turning point, change began to dominate thereafter. Moreover, each succeeding culture contained a larger array of finer implements than the last. Bone, antler, and ivory became increasingly important raw materials for tool making, particularly for fine, flexible and sharp implements. And, most striking of all, there began to emerge a previously unseen degree of variability in the form of tool kits found in neighboring sites, a variability that has been explained variously as discrete functional differentiation or cultural expression through style.

The major post-Acheulian tool culture was the Mousterian, associated with the origin of archaic *Homo sapiens:* this was known in Europe as the Middle Paleolithic. The Mousterian continued through to around 35,000 to 40,000 years ago, which coincided with the appearance of fully modern humans, *Homo sapiens sapiens* in Europe. The 100,000 year tenure of the Mousterian was overtaken by a succession of tool kits, known as the Upper Paleolithic, that displayed the ever increasing virtuosity of the tool makers, so much so that some implements lost any function they might have had and instead assumed some kind of abstract symbolism. (In Africa the equivalent periods to the Middle Paleolithic and Upper Paleolithic are called the Middle Stone Age and the Later Stone Age.)

Towards the end of the Acheulian era, some 200,000 years ago, there arose in South Africa a new technique for the production of large flakes that was to foreshadow later developments in tool technology. Known as the Levallois technique, this new development involved a much more intensive preparation of a core than had hitherto been the practice. Virtually complete flakes could then be struck from the core at a single blow, although they were typically retouched to give the final desired shape. It was principally a refinement and development of the Levallois technique that formed the basis of Mousterian tool technology of the Middle Paleolithic.

One immediate practical consequence of careful core preparation is a greater efficiency in the use of raw materials. For example, the basic Acheulian method yielded just 5.1 to 20.3 centimeters of cutting edge from 0.45 kilogram of flint, whereas a Mousterian tool maker could strike 10.2 meters of cutting edge from the same amount of starting material. This trajectory of greater efficiency soared following the origin of modern humans who could manufacture 12 meters of cutting edge from 0.45 kilogram of flint, struck in the form of long sharp blades.

In addition to a greater efficiency in the use of raw materials, the Mousterians also made a much wider range of stone implements by patiently and sensitively retouching the basic flakes. The great French archeologist Francois Bordes counted as many as 60 systematically produced categories across the whole range of the Mousterian, although at no single site would all 60 types be found. The Mousterian tool kit included small handaxes, flake blades, scrapers, and many items for detailed work, such as points, awls and burins.

Neanderthal people, *Homo sapiens neanderthalensis*, who have been particularly associated with the Mousterian culture, occupied the Old World from western Europe through to central Asia. Because of certain characteristic patterns associated with some of the Mousterian tool kits, Bordes suggested that there were several different tribes of Neanderthals, each expressing their social identity through stylistic

Middle Paleolithic artifacts: These are typically retouched flakes of various types, made between 200,000 and 40,000 years ago. Top row, left to right: Mousterian point; Levallois point; Levallois flake (tortoise); Levallois core; disc core. Bottom row, left to right: Mousterian handaxe; single convex side scraper; Quina scraper; limace; denticulate. (Scale bar: 5 cm.) (Courtesy of Roger Lewin and Bruce Bradley.)

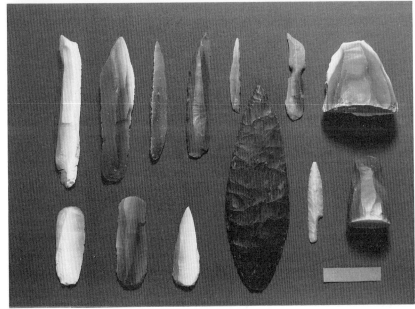

Upper Paleolithic artifacts: These are typically formed from retouched blades and are therefore finer than Middle Paleolithic tools. Top row, left to right: burin on a truncated blade; dihedral burin; gravette point; backed knife; backed bladelet; strangulated blade; blade core. Bottom row, left to right: end scraper; double end scraper; end scraper/dihedral burin; Solutrean laurel leaf blade; Solutrean shouldered point; prismatic blade core. (Scale bar: 5 cm.) (Courtesy of Roger Lewin and Bruce Bradley.)

conformity. Another interpretation of the very real differences between some of the sites, advanced particularly by American archeologists Lewis and Sally Binford, is that they represent different sets of tools required for different functions.

It would be surprising indeed if, for example, the implements employed for woodworking were not different from those used at a butchery site, so functional differentiation must play a part in the variation seen in Mousterian archeological sites. However, it would be equally surprising if the Mousterian people, as intelligent and sensitive as they evidently were, did not perceive complex social structures that were differentiated through identifiable styles, in tool making and no doubt in other ways too.

In many ways, the shift 40,000 years ago from the Middle Paleolithic to the Upper Paleolithic was not as dramatic as that between the Acheulians and the Mousterians. Fine narrow blades were a prominent feature of the Upper Paleolithic tool kits, as were delicate implements of bone and antler. Tool preparation involved a greater emphasis on precisely controlled pressure flaking as opposed to a free stroke with a hammerstone. These items had been part of the Middle Paleolithic, but they were

not especially emphasized. A key innovation of Upper Paleolithic technologies, however, was the use of bone and ivory as raw material, often in the production of fine, sharp implements.

Archeologists recognize new tools in the technologies of these newly evolved modern humans, so that overall there were as many as 100 identifiable implements in the Upper Paleolithic. These included hafted implements, which involved the conceptual and technological advance of combining two artifacts. The important point, however, is that new items arose every few thousand years throughout the Upper Paleolithic as compared with every 5 or 10 thousand in the earlier era.

The tremendous emphasis of research on the Upper Paleolithic in Europe has yielded a wealth of detail as compared with the rest of the world. As a result certain 'cultures' have been named as representing specific periods within the Upper Paleolithic. For example, two principal cultures, the Aurignacian and the Perigordian, are described as having coexisted for much of the earliest part of the period, from around 35,000 years to around 20,000 years ago. As one would anticipate, there is considerable consistent variation within these cultures too, which has again engendered discus-

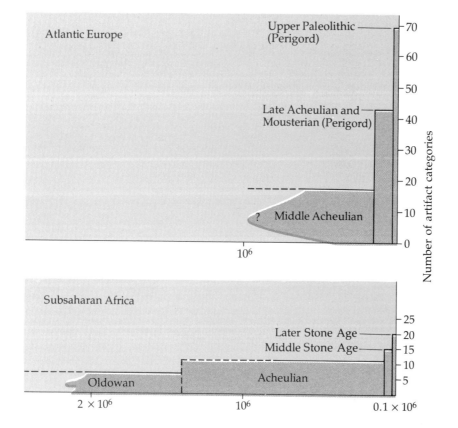

Increase in pace of new tool techologies: The emergence of new distinct tool technologies at an ever-accelerating pace through time was accompanied by a great increase in the number of distinct implements comprising the cultures.

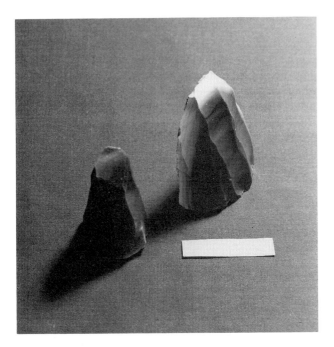

Prismatic blade core (left) and blade core (right): Two sources of narrow flint blades in Upper Paleolithic tool technology. Repeated blows around the edge of the flat surface (upon which the cores are resting in the picture) yields large numbers of sharp blades which can be finely retouched. (Scale bar: 5 cm.) (Courtesy of Roger Lewin and Bruce Bradley.)

sions on cultural style and utilitarian need. The successors to the Perigordians were the Solutreans, who, among other things, were distinguished by their fine laurel leaf blades and their invention of the eyed sewing needle. The Solutreans, 20,000 to 16,000 years ago, were followed by the Magdalenians, who were responsible for the high art of Ice Age Europe.

With the passing of the Ice Age 10,000 years ago the Magdalenian culture passed too, to be replaced by an ever greater burgeoning of traditions associated with the new activities of the Agricultural Revolution. A new archeology began to arise, the archeology of towns and cities, of trade and of conflict, patterns that in a remarkably short period of time were to be found throughout the world, both Old and New.

Key questions:

● How important has Eurocentrism in archeology been in interpretations of culture since the origin of modern humans?
● What kinds of evidence are required to address the issue of different cultures versus different functions at, for instance, Perigoridan and Aurignacian sites?
● What is the significance of the introduction of bone as a raw material into Upper Paleolithic and Later Stone Age cultures?
● Human ancestors became anatomically modern long before they developed modern tool technologies: what is the significance of this uncoupling?

Key references:

D. Bordes, 'The Upper Paleolithic: c33,000–10,000 BC', in *France before the Romans*, edited by Stewart Piggott, Glynn Daniel, and Charles McBurney, Thames and Hudson, 1972, pp 30–60.
F. Bordes, 'Mousterian cultures in France', *Science*, vol 134, pp 803–810 (1861).
Richard Klein, *Ice-age hunters of the Ukraine*, The University of Chicago Press, 1973.
Richard Klein, 'The Stone Age prehistory of Southern Africa', *Annual Reviews of Anthropology*, vol 12, pp 25–48 (1983).

26 / Art in prehistory

Traditionally, the study of prehistoric art has meant the study of prehistoric art in Europe, specifically in southwest France and northern Spain during the period 35,000 to 10,000 years ago. There is no doubt that artistic expression flowed elsewhere in the Old World at this time—most notably in East Africa—but accidents of history and preservation have endowed Europe with a rich record of painted, engraved and carved images that, properly interpreted, must give some insight into the workings of the human mind at this point in our history.

The beginning of European prehistoric art apparently coincides with the arrival there of anatomically modern humans. The continent at the time was in the grip of the last great Pleistocene glaciation, which began 75,000 years ago and ended 10,000 years ago. After their arrival in western Europe, modern human populations would eventually have to endure the glacial maximum, about 18,000 years ago, which happens to coincide with what many scholars consider to be the high point of prehistoric art.

For much of the last glaciation, northern Europe lay buried beneath an ice sheet more than a kilometer thick in many places. To the south of the ice, the climate was colder and drier than in interglacial times, with open grassy plains replacing dense woodland and forest that once mantled much of Europe. Regions with topographical relief experienced more variety of vegetation cover, patches of woodland surviving in sheltered valleys. Herds of horses, bison, and aurochs (forerunners of today's cattle) roamed the plains of France and Spain, as did reindeer and ibex. Woolly mammoth and rhinoceros were to be seen too, although they were more common further north and east.

Within the glacial period, the climate fluctuated in severity, the animal and plant communities fluctuating with it. Warmer climes brought a return of woodland, and with it came woodland creatures, such as wild boar. Gone were the animals of the

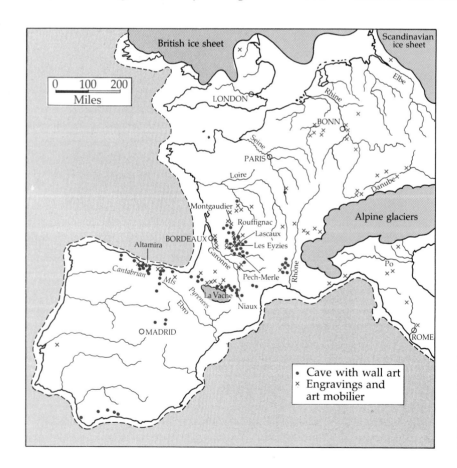

Distribution of art sites in Europe: The limestone caves of Ice Age Europe have preserved a rich legacy of Paleolithic art. Although there was a certain stylistic continuity in cave painting, motifs in art mobilier displayed much more variability.

(a)

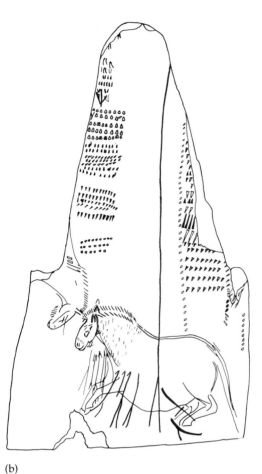

(b)

Examples of cave art:

(a) Fragment of reindeer antler from La Marche, France, around 12,000 years old. Apparently used as an implement for shaping flint tools, the antler fragment is engraved with a pregnant mare, which seems to have been symbolically killed by a series of engraved arrows. Above the horse is a set of notches made at different times by different tools. Marshack interprets the marks as a notation series, perhaps documenting the passing lunar cycles.

(b) A drawing of the engraving 'unrolled'.

Below

(e) Vogelherd horse, carved from mammoth ivory about 30,000 years ago and worn smooth by frequent handling over a long period of time. The horse, which is the oldest animal carving known, measures 5 centimeters.

(f) The black outline of this horse was painted on the wall of a cave, Peche-Merle, in France 15,000 years ago. infrared analysis indicates that the mixture of black and red dots were added over a period of time. The black hand stencils are also later additions. Does the Peche-Merle horse, one of two in the cave, indicate 'use' of art? (Courtesy of Alexander Marshack.)

(c)

(d)

(e)

(f)

(c) Engraved antler baton from Montgaudier, France, dated about 10,000 years old. Perhaps used in straightening the shafts of arrows or even spears, the collection of items engraved on the baton suggests a representation of spring.

(d) A drawing of the engraved antler-baton 'unrolled'.

plains, temporarily at least. Sometimes the fluctuations extended over many millennia, sometimes within dramatically short periods: for instance, one region in southern France went from open grassland to oak forest and back to grassland, all in the space of a few hundred years. The artists of the Ice Age lived in changing times.

A brief survey of Upper Paleolithic art yields a few generalities.

For instance, the painted wall art is mainly of large mammals—bison, aurochs, deer, horses, mammoth, ibex, and so on—but carnivores are rare. Birds, plants, and humans are only infrequently represented, and the latter often quite schematically when they do appear. The painted images are often very good, naturalistic representations of single animals, or small groups of individuals, but there is little sense of natural scenes.

Engraved or carved images, particularly on portable objects such as spear throwers, batons, pendants, and blade punches, often have more detail in their execution, and overall give a sense of a wider representation of nature, including the large mammals seen in wall art (although in different proportions). For instance, birds, fish, and plants are often depicted, sometimes in rich combination: but again, this seems not to be the representation of a scene so much as an idea, such as a season. Interestingly, in body ornamentation such as necklaces and pendants, carnivore teeth are present in very high proportion, a striking contrast to the wall art.

The human image occurs more frequently as carved and engraved images than in painting, but again they are often schematic, as in the famous 'Venuses'. From one site, however, there was recovered a cache of more than 200 small engraved human faces, completely lifelike and individualistic—a portrait gallery from 20,000 years ago.

When the Ice Age finally came to an end, so did the art, at least in the generally naturalistic representational style that had persisted for 25,000 years: geometric patterns became predominant, and people apparently no longer sought out deep caves in which to paint.

The overriding sense with both wall and portable images is of art that is *used*, perhaps with portable objects being more personal than painted or engraved images on walls. The task of the scholar, says Margaret Conkey of the University of California, Berkely, is not to wonder what it all means but to ask: 'what was the social context of the art that made it meaningful to the people who made and used these images?'

When, at the turn of the century, Upper Paleolithic art was accepted as being genuinely of great antiquity it was interpreted merely as 'art for art's sake', an idea that has recently been revived. For instance, John Halverson of the University of California at Santa Cruz interprets the directness and simplicity of Paleolithic images as basic artistic expression 'unmediated by cognitive reflection'. What we see in the art, he suggests, is the product 'not of "primitive mind" but "primal mind", human consciousness in the process of growth'. Most scholars, however, believe there was more maturity

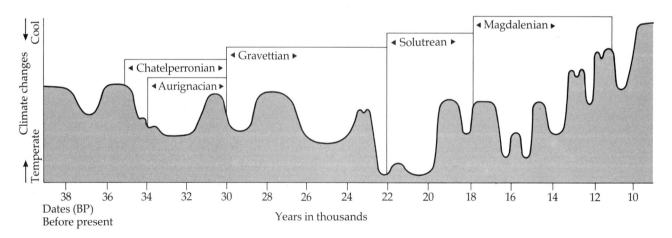

Climatic fluctuations: Although European prehistoric art was the product of the Ice Age, temperatures did fluctuate somewhat throughout this period, driving dramatic shifts in ecological patterns. The most frigid period, from 22,000 to about 18,000 years ago, preceded the high point of prehistoric art, the Magdalenian. (Courtesy of the Randall White/American Museum of Natural History.)

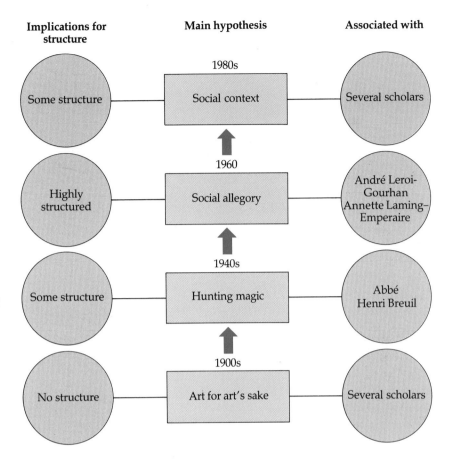

Implications for structure	Main hypothesis	Associated with

1980s

Some structure — Social context — Several scholars

1960

Highly structured — Social allegory — André Leroi-Gourhan Annette Laming-Emperaire

1940s

Some structure — Hunting magic — Abbé Henri Breuil

1900s

No structure — Art for art's sake — Several scholars

Changing theories: After cave and portable art was finally accepted (in the late 1890s) as genuine products of ancient people, scholars' interpretations of its meaning have evolved through different stages. Shown here are the four major hypotheses, with dates and main proponents (where applicable). The different hypotheses had different implications for how the art was distributed—structured—within the caves.

and complexity within human consciousness and its products in the Upper Paleolithic.

The first systematic study of Ice Age art was by the great French archeologist, the Abbe Henri Breuil, who throughout the first half of this century carefully copied images from many sites, and attempted a chronology based on artistic style. Eventually he concluded that the art served the purpose of 'hunting magic', a way of ensuring fruitful hunts and propitiating the victims. There are good ethnographic parallels for such an interpretation.

Breuil's hunting magic explanation was replaced in the 1960s by the notion that the art somehow reflected the society that produced it, a thesis developed independently by French archeologists Andre Leroi-Gourhan and Annette Laming-Emperaire. They noted that the inventory of animals depicted was comparable throughout Europe and considered that the presentation was remarkably stable through time, an observation that contrasts with the much more locally idiosyncratic nature of portable art.

Wall art, for Leroi-Gourhan and Laming-Emperaire, reflected the duality of maleness and femaleness in society. Certain images were said to represent maleness while others were female. The arrangement of the images within caves was such that female representations were at the center, with male representation around the periphery, thereby reflecting a certain type of social structure. Although the two researchers did not fully agree on which images represented maleness and which femaleness, their work had the important effect of emphasizing social context in interpreting Paleolithic art.

So, where Breuil's explanation required no overall structure of the images within the caves, Leroi-Gourhan and Laming-Emperaire's very clearly did. Both explanations, however, were essentially monolithic. In recent years this too is changing. 'We are beginning to see a great deal more diversity and complexity in Upper Paleolithic art,' explains Randall White of New York University, 'and this affects the way we envisage what was going on during this important stage of human evolution.'

The Upper Paleolithic is divided into different cultural periods, based upon the tool technologies of the time (see unit 25). And throughout these different cultures different aspects of the art changed in various ways, which is what Breuil had tried to base his chronology upon. 'It is important not to get the idea this pattern of change advanced on a broad front', cautions White. 'In addition to differences through time, there are differences between regions, real geographic variations.' These spatial and temporal variation in tool cultures are matched by similar variations in the art, but there is no precise correlation between a culture's technology and its art. It is therefore clear that there can be no monolithic explanation of the meaning of the art.

Hunting magic may well be the explanation for some of the images. But ritual of other kinds almost certainly centered of the art too: something other than practicality drove Upper Paleolithic people to seek out and decorate deep caves, which appear to be otherwise unutilized. Still another possible role of the art is in reflecting society as a whole, as Leroi-Gourhan and Laming-Emperaire had suggested. For instance, the famous cave of Altimira in northern Spain has a circle of painted bison, surrounded by other animals, as the centerpiece of its ceiling. According to archeological and other evidence, Altimira appears to have been the focus of seasonal aggregation of people from several different geographical regions, a phenomenon that occurs in modern hunter–gatherers. Conkey, who has studied the wall images and portable art at Altimira, speculates that the circle of bison represents this process of aggregation.

Many portable art objects are decorated with geometric patterns; some have pictures of animals, fish, and plants; and others have series of seemingly random notches. Alexander Marshack, an independent researcher, performed detailed studies of such objects, and suggested that, for instance, some of the combinations of images might represent seasons of the year: the images of a male and a female seal, a male salmon, two coiled snakes, and a flower in bloom, all engraved on a reindeer antler baton is one such example.

Marshack has also proposed that some of the series of dots and lines on various objects might represent some kind of notation—the lunar cycle, for instance. He also has evidence that painted and engraved images were sometimes worked on repeatedly, being constantly added to or modified. This, he suggests, implies images being *used*, not simply being made.

In recent independent investigations, Denis Vialou, of the Musée de l'Homme in Paris, and Henri Delport, of the Musée des Antiquites Nationales, near Paris, conclude that there is less overall uniformity of structure among the painted caves than originally envisaged by Leroi-Gourhan and Laming-Emperaire. They acknowledge that there is some kind of structure within most of the caves, but caution that each cave should be viewed as a separate expression.

Diversity, then, begins to emerge as a more realistic interpretative lens through which to view the Upper Paleolithic, a diversity of people, a diversity of cultures, and a diversity of the art. And there is a shift from trying to understand what an individual image or set of images might mean to how one might understand the social context in which those images were produced. Most of all, there is an attempt to try to divest modern interpretations of the bias inherent in modern eyes and minds. As Conkey says: 'Perhaps we have closed off certain lines of inquiry, simply by using the label "art".'

Key questions:

• In what ways are modern interpretations of Paleolithic art most likely to be biased?
• How would one test the hypothesis that, in some cases at least, Paleolithic art is a form of hunting magic?
• What possible interpretations are there for the relative rarity of carnivore images in wall art as against the extensive use of carnivore teeth in body ornamentation?
• Can the art of another culture ever be completely understood by those outside it?

Key references:

Margaret Conkey, 'On the origins of paleolithic art', *British Archeological Reviews*, vol 164, pp 201–227 (1983).
Whitney Davis, 'The origins of image making', *Current Anthropology*, vol 27, pp 193–215 (1986).
John Halverson, 'Art for art's sake in the paleolithic', *Current Anthropology*, vol 28, pp 63–89 (1987).
Arlette Leroi-Gourhan, 'The archeology of Lascaux Cave', *Scientific American*, June 1982, pp 104–112.
Andre Leroi-Gourhan, 'The evolution of paleolithic art', *Scientific American*, February 1968.
J.D. Lewis-Williams, 'Cognitive and optical illusions in San rock art research', *Current Anthropology*, vol 27, pp 171–177 (1986).
Randall White, *Dark caves, bright visions*, The American Museum of Natural History, 1986.

27 / The expanded brain

This unit focuses on the context and consequences of human brain evolution. The context will expose some of the biological constraints under which this extraordinary evolutionary event occurred. And the consequences should help illuminate some of the selection pressures which fueled that event.

The question to be answered is: how did humans come to possess such extraordinary powers of creative intelligence, powers that surely outstrip what would have been necessary in the practical day to day life of technologically primitive hunter—gatherers? In exploring this conundrum we will cover three aspects of human mental evolution: the expansion of the brain and intelligence, consciousness, and language (unit 28). Inevitably, these three quality are tightly intertwined.

Although the hominid lineage stretches back some 6 million years, fossil evidence for brain size goes back only 3 million years, to the specimens of *Australopithecus afarensis* at the Hadar, Ethiopia. Given this somewhat limited view of hominid history, it is nevertheless apparent that brain expansion has been great and rapid: a threefold increase occurred in that 3 million years, going from around 400 cm^3 to 1350 cm^3, the average of modern populations. Impressive in itself, it is the more so when, as pointed out by Harry Jerison of the University of California, Los Angles, 'there is no evidence of a change in any other mammals in [this same period]'. In other words, brain expansion among hominids was not just part of a recent, general mammalian pattern.

To understand this expansion more fully we will look first at some of the characteristics of the human brain in the context of primate biology; we will then turn to the fossil evidence of this expansion; finally, we will consider some of the ideas currently offered to explain the phenomenon.

First of all, the brain is a very expensive organ to maintain. In adult humans, for instance, even though it represents just 2 per cent of the total body weight, the brain consumes some 18 per cent of the energy budget. 'One might therefore ask,' says Robert Martin, 'how, in the course of human evolution, additional energy was made progressively

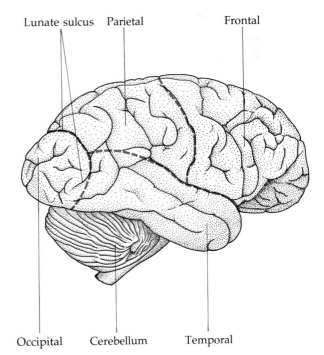

Parietal   Frontal

Lunate sulcus   Parietal   Frontal

Occipital Cerebellum   Temporal Broca's area

Occipital   Cerebellum   Temporal

Diagram of the typical ape and human brain pattern: The large human brain (left) compared with that of a chimpanzee is also distinguished by its relatively small occipital lobe and large parietal lobe. (Courtesy of Ralph Holloway/*Scientific American*, 1974, all rights reserved.)

available to meet the needs of an ever-increasing brain size.'

As we saw in unit 10, life-history factors—gestation length, metabolic rate, precociality versus altriciality, and so on—have important bearings on the size of brain a species can develop. In this context, two major ideas have been advanced in recent years that bear on the special problem faced by hominids in underpinning brain expansion.

The first, proposed by Martin, is that the mother's metabolic rate is key to the size of brain a species can afford: the higher the metabolic rate, the bigger the potential brain size. The second, proposed by Mark Pagel and Paul Harvey of Oxford University, is that gestation time and litter size are the determining factors: long gestation, with a litter of one is optimal for a large-brained species. Both hypotheses are said by their authors to have empirical support, but a debate continues as to which is the more germane. Whichever case proves to be correct, both pathways require the same kind of environmental context: stable, high-energy food supply, with minimum predation pressure.

In being well endowed mentally, humans and other primates are a part of a very clear pattern among vertebrates as a whole. Depending somewhat on the measure you use, mammals are about 10 times brainier than reptiles, and in their turn, reptiles are about 10 times brainier than amphibians. Underlying this stepwise progression, which represents successive major evolutionary innovations and radiations, is the building of more and more sophisticated 'reality' in species' heads.

So, being mammals, primates by definition are better equipped mentally than any reptile. However, two orders of mammal have significantly larger brains than the rest: they are Primates and Cetaceans (toothed whales). And among primates, the anthropoids (monkeys and apes) are brainier still. Only humans lie off that monkey/ape axis: the brain of *Homo sapiens* is three times bigger than that of an ape of the same body size.

The need to grow such a large brain has distorted several basic life-history characteristics seen in other primates. For instance, the adult ape brain is about 2.3 times bigger than the brain in the newborn (neonate); in humans the difference is 3.5 times. More dramatic, however, is the size of the human neonate compared with ape newborns. Even though humans are of similar body size to apes (57 kilograms, compared to 30 to 100 kilograms) and have a similar gestation period (270 days,

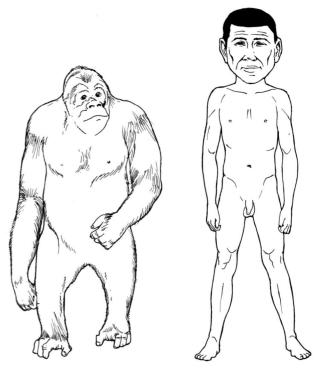

Expanded human brain: The human brain is three times bigger than an ape's brain would be, given the same body size.

compared to 245 to 270 days), human neonates are approximately twice as big and have brains twice as big as ape newborns. 'From this it can be concluded that human mothers devote a relatively greater quantity of energy and other resources to fetal brain and body development over a standard time than do our closest relative, the great apes', notes Martin.

Another major difference is the pattern of growth. In mammals with precocial young—which includes primates—brain growth proceeds rapidly until birth, whereupon a slower phase ensues for about a year. In humans, the prenatal phase of rapid brain growth continues until well after birth, a pattern that is seen in altricial species. One difference with altricial species, however, is that in humans the rapid postnatal phase (fetal rate) brain growth continues for a relatively longer period than is typical of such species. The effect is to give humans the equivalent of a 21 month gestation period (9 in the uterus, 12 outside).

This unique pattern of development has been called secondary altriciality. One important consequence is that human infants are far more helpless and for a much longer time than the young of the great apes. This extended period of infant care

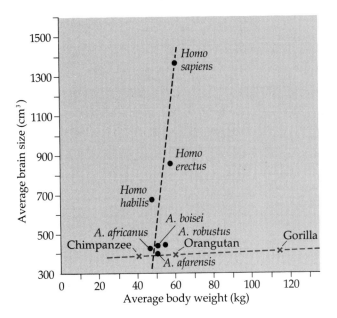

Brains and bodies: Even though there was not a dramatic increase in body size in the *Homo* lineage, absolute (and therefore relative) brain size expanded significantly from *habilis* to *erectus* to *sapiens*. Brain size did not alter significantly among the australopithecines, nor among the modern apes, despite a large body size difference in the latter.

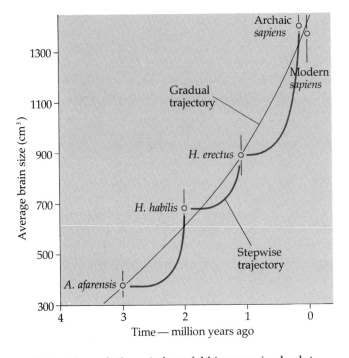

Brains through time: A three-fold increase in absolute brain size occurred during the past 3 million years. Whether this increase took place gradually (as indicated by the smooth slope) or episodically (as indicated by the steps) is a matter that will be settled only with more fossils, accurately identified.

and of subsequent 'schooling' must have had a major impact on the social life of hominids.

Fossil evidence of brain evolution is of two types: an indication of absolute size, and information about the surface features—convolutions and fissures—of the brain. Both pieces of evidence can be obtained from either natural or artificial endocasts, which show the convolutions of the brain as they had become impressed on the inner surface of the cranium.

Brain size is the first and most obvious piece of information to be gleaned, and this can often be gained even with partial crania. There is therefore a fair amount of data about the expansion of brain size, beginning with *Australopithecus afarensis*, a little more than 3 million years ago. Measured at a little less than 400 cm^3, this early australopithecine brain is often said to be about the same as that of modern gorilla and chimpanzee brains. However, this is misleading, for two reasons: first, early australopithecines were smaller in body size than modern gorillas; second, modern ape brains almost certainly are expanded over what their 3 million-year-old ancestors' brains would have been. It is therefore safe to say that brain expansion had already been established by the time *Australopithecus afarensis* appeared.

In bold terms, then, brain size for the australopithecines was close to 400 cm^3, and it never really increased throughout the tenure of this genus. More marked expansion is seen with the origin of the genus *Homo*, specifically *Homo habilis*, which existed from about 2.5 million to 1.6 million years ago and had a range of brain size of about 650 to 800 cm^3. The range for *Homo erectus*, 1.6 million to about 300,000 years ago, is 850 to a little more than 1000 cm^3. Archaic *Homo sapiens*, including Neanderthals range from 1100 to more than 1400 cm^3, which is larger than in modern humans.

By looking at overall brain structure as revealed in endocasts, it is possible to differentiate between an apelike and a humanlike brain organization. The four lobes in each hemisphere are as follows: the frontal, the temporal, the parietal, and the occipital. Very briefly, a brain in which the parietal and temporal lobes predominate is humanlike, whereas in apelike brains these areas are much smaller.

Ralph Holloway of Columbia University has examined in detail a wide range of hominid fossil endocasts, including *Australopithecus afarensis*. The conclusion was a surprise: the brain had apparently been reorganized into a humanlike configuration right from these earliest times of the hominid lineage. Even though it is not possible to see clearly the humanity in the humanlike brain organization, this result implied that something special was happening very early on in hominid evolution.

Just recently Holloway's conclusions have been challenged by Dean Falk of Purdue University. Although she agrees with Holloway that brain of *Homo* species are reorganized in the human direction, she contends that australopithecine brains are essentially apelike. The precise location of some of the key fissures and divisions between lobes is often very difficult and open to interpretation. In this case the differences of opinion continue unresolved.

It is relatively easy to plot brain expansion through hominid history, but how are we to measure the rise of intelligence through time? The archeological record is notoriously sparse with tangible indications of the working of the mind: tales told around a camp fire, complex sand drawings, dances embodying social mythology leave no trace, and yet are the essence of humanity in a hunter–gatherer society. Yes, paintings and engravings betoken mental activities beyond basic subsistence, something we can identify as quintessentially human, but these come very late in our history (unit 26).

What we are left with are the stone tools and other clues to economic activity. As we saw in units 24 and 25, the imposition of standardization and expansion of complexity were very slow to develop in prehistoric stone-tool industries. Using the criteria of psychological theory (that of Piaget), archeologist Thomas Wynn of the University of Colorado has analyzed some of the early stone-tool industries, looking for signs of humanlike intelligence. 'The evolution of a uniquely hominid intelligence had not occurred by Oldowan times', he concludes, referring to the fossil and archeological remains at Olduvai Gorge, dating between 1.9 and 1.6 million years. This was the time of *Homo habilis* but prior to *Homo erectus*. 'This suggests that selection for a complex organizing intelligence was not part of the original hominid adaptation.'

One last insight that fossil evidence might allow in relation to expanding brain size concerns its impact on social organization, specifically in infant care. Once hominids shifted from the basic primate pattern of brain growth, producing instead a much more helpless infant whose brain continued to grow at the fetal rate, then great allocation of time and

The social milieu: Socializing has become an important part of primate life: making alliances, and exploiting knowledge of others' alliances, is key to an individual's reproductive success. Biologists are coming to believe that the intellectual demands of complex social interaction was an important force of natural selection in the expansion of primate—and ultimately, human—brains.

resources would be needed for rearing offspring. It is theoretically possible, as Robert Martin points out, that no change in infant care would be needed until after the adult human brain size exceeded 873 cm^3, which is the transition size between *Homo habilis* and *Homo erectus*. The argument is as follows.

Suppose that hominids had been able to make all the other changes in fetal development—speeding up body and brain growth—but then reverted to the basic primate pattern in the neonate. This pattern would have allowed for the doubling (actually, × 2.3) of brain growth at birth. Now, the brain size of human infants is 384 cm^3: multiply this by 2.3 and you get 873 cm^3. This theoretical calculation depends on the assumption that the birth canal in the pelvis of *Homo habilis* or early *Homo erectus* females would be able to accommodate a neonate's head the size of a modern infant's.

From the fossil evidence available so far, it is clear that the hominid birth canal was smaller than the modern female's at this point in our history. Which means that a shift to humanlike postnatal brain growth patterns would have had to have occurred already in *Homo habilis*, presumably with the concomitant impact on social organization.

From fossils we turn to theories; theories about the selection pressure (or pressures) that powered hominid brain expansion. Popular for a very long time was the notion that the very obvious difference between hominids and apes—that humans made and used stone tools—was the most likely cause: the tripling of hominid brain size was accompanied by an ever increasing complexity of tool technology. 'Man the tool maker' was the encapsulation of this approach in the 1950s, to be followed a decade later by 'Man the hunter'. In either case, the emphasis was on the mastering of practical affairs as the engine of hominid brain expansion.

In more recent times new ideas have emerged, which might be encapsulated in the phrase 'Man the social animal'. Part of the reason for this shift of opinion has come from primate field studies, which are now reaching an important point of maturity. In addition, there has been a greater introspection about the human mind itself, particularly consciousness.

The new insight into 'Man the social animal' begins with a paradox, similar in nature to the human paradox: laboratory tests have demonstrated beyond doubt that monkeys and apes are extraordinarily intelligent, and yet field studies have revealed that the daily lives of these creatures is relatively undemanding, in the realm of subsistence at least. Why, then, did this high degree of intelligence develop?

The answer may lie in the realm of primate social life. Although, superficially, a primate's social environment does not appear to be more demanding than that of other mammals—the size and composition of social groups is matched among antelope

species, for example—the *interactions* within the group are far more complex. In other words, for a nonhuman primate in the field, learning the distribution and probable time of ripening of food sources in the environment is intellectual child's play compared with predicting—and manipulating—the behavior of other individuals in the group. But why should social interactions be so complex, so Machiavellian in primate societies?

When you observe other mammal species and see instances of conflict between two individuals it is usually easy to predict which will triumph: the larger one, or the one with bigger canines or bigger antlers, or whatever is the appropriate weapon for combat. Not so in monkeys and apes. Individuals spend a lot of time establishing networks of 'friendships' and observing the alliances of others. As a result, a physically inferior individual can triumph over a stronger individual, provided the challenge is timed so that friends are at hand to help the challenger and while the victim's allies are absent.

'Alliances are far more complex social interactions than are two-animal contests', says Alexander Harcourt of Cambridge University. 'The information processing abilities required for success are far greater: complexity is geometrically, not arithmetically, increased with the addition of further participants in an interactions....In sum, primates are consumate social tacticians.'

In a recent survey of much of the field data relevant to primate social intelligence, Dorothy Cheney, Robert Seyfarth (both of the University of Pennsylvania), and Barbara Smutts (of the University of Michigan) posed the question: 'are [primates] capable of some of the higher cognitive processes

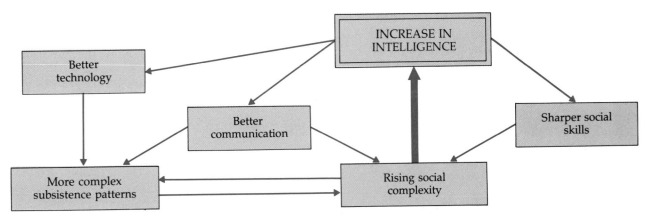

Social complexity and increased intelligence: The need to cope with rising social complexity—including increasingly demanding subsistence patterns but, particularly, a more ramified social structure and unpredictable social interactions—may have been a keen selection pressure for increased intelligence.

that are central to human social interactions?' The question is important, because if anthropoid intellect, honed by complex social interaction, is merely sharper than the average mammal's and more adept at solving psychologist's puzzles, then it doesn't match the *creative* intelligence in which we are interested.

Cheney and her colleagues had no difficulty in finding many examples of primate behavior that appear to reflect humanlike social cognition. The researchers conclude that 'primates can predict the consequences of their behavior for others and they understand enough about the motives of others to be able to be capable of deceit and other subtle forms of manipulation'. So, if, as seems to be the case, nonhuman primate intellect has been honed, not in the realm of practical affairs but in the hard school of social interaction, one is still left with the question: why? Why have primates found it advantageous to indulge in alliance building and manipulation? The answer, again from field studies, is that individuals who are adept at building and maintaining alliances are also reproductively more successful: making alliances aids in access to potential mating opportunities.

Once a lineage takes the evolutionary step of using social alliances to bolster reproductive success, it finds itself in what Nicholas Humphrey, a Cambridge University psychologist, calls an evolutionary ratchet. 'Once a society has reached a certain level of complexity, then new internal pressures must arise which act to increase its complexity still further,' he explains, 'for, in a society [of this kind], an animal's intellectual "adversaries" are members of his own breeding community. And in these circumstances there can be no going back.'

And where in all this does consciousness fit in? Humphrey describes it as an 'inner eye', with pun intended. Consciousness is a tool—the ultimate tool—of the social animal. Through being able to look into one's own mind and 'see' its reactions to things and other individuals, one can more precisely predict how others will react to those same things and individuals. Consciousness builds a better reality, one that is attuned to the highly social world humans inhabit.

Key questions:

- What limitations are there in measuring differences in intelligence from differences in brain size and overall organization?
- How might one infer levels of intelligence from different stone-tool technologies?
- What key pieces of information might lend support to the 'Man the social animal' hypothesis?
- How would one test whether nonhuman primates possessed a humanlike consciousness?

Key references:

D. Cheney et al., 'Social relationships and social cognition', Science, pp 1361–1366 (1986).

D. Falk, 'The petrified brain', Natural History, September 1984, pp 36–39.

A.H. Harcourt, 'Alliances in contests and social intelligence', in Social expertise and the evolution of intellect, edited by R. Byrne and A. Witen, Oxford University Press, 1988.

R.L. Holloway, 'Human brain evolution', Canadian Journal of Anthropology, vol 3, pp 215–230 (1983).

N.K. Humphrey, 'The social function of intellect', in Growing points in ethology, edited by P.P.G Bateson and R.A. Hinde, Cambridge University Press, 1976.

N.K. Humphrey, The inner eye, Faber and Faber, 1986.

R.D. Martin, 'Human brain evolution in an ecological context', Fifty-second James Arthur Lecture, American Museum of Natural History, 1983.

M.D. Pagel and P.H. Harvey, 'How mammals produce large-brained offspring', Evolution, pp 948–957 (1988).

T. Wynn, 'The intelligence of Oldowan hominids', Journal of Human Evolution, pp 529–541 (1981).

28 / Origin of language

One great frustration for anthropologists is that language—this species-specific quality with which we are so justifiably impressed—is virtually invisible in the archeological record. Not until permanent forms of writing were used can one be certain that language existed. The first encryptions in clay tablets are found among relics of the Sumarian civilization of some 6000 years ago, but no one would argue that this marks the origin of the spoken word. No, other clues have to be sought: in stone tools, among indications of social and economic organization, in the content and context of paintings and other forms of artistic expression, and in the fossil remains themselves.

First, the fossils. In recent years researchers have pursued two kinds of evidence from fossil hominids: first, information to be gleaned from endocasts, those crude maps of the surface features of the brain; second, indications of the structure of the voice-producing structures in the neck, the larynx and pharynx.

'It is the bias of this writer that the origins of human language behavior extended rather far back into the paleontological past,' suggests Ralph Holloway, 'and were nascent, but growing, during australopithecine times of roughly 2.5 to 3.5 million years ago.' Holloway bases his bias—and he is explicit in saying it is a bias, not a firm conclusion—on humanlike reorganization which he says he can detect in even the earliest hominids (unit 27). More than general architecture, however, is the presence of a particular language-associated feature on the general cerebral landscape. This feature, Broca's area, is seen as a small lump on the left side of the brain, towards the front: present in all modern humans, it is also to be seen in the earliest *Homo*, but not *Australopithecus*, species.

The major neural machinery for language functions is located in the left hemisphere in the great majority of modern humans, even in most left-handed people. But, as with many complex mental functions, language capabilities cannot be pinpointed precisely to particular centers. It is true, as mentioned above, that Broca's area is associated with language, particularly with the production of sound. And a second center, Wernicke's area, located somewhat behind Broca's area, is involved in the perception of sound. But many aspects of language, for instance the lexicon, or vocabulary with which we work, defy precise localization.

If the fossil brains give tantalizing hints of verbal

Language centers: Wernicke's area, which appears to be responsible for content and comprehension of speech, is connected by a nerve bundle called the arcuate fasciculus to Broca's area, which controls the muscles of the lips, jaw, tongue, soft palate, and vocal cords during speech. These language centers are usually located in the left cerebral hemisphere, including many left-handed people. (Courtesy of Norman Geschwind/*Scientific American*, 1972, all rights reserved.)

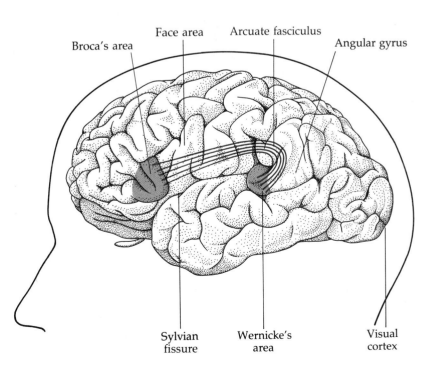

Face area · Arcuate fasciculus

Broca's area · Angular gyrus

Sylvian fissure · Wernicke's area · Visual cortex

skills in our ancestors, what can we learn from the voice-producing apparatus? A number of researchers have pursued this question in recent years—particularly Edmund Crelin, Philip Lieberman, and Jeffrey

Laitman. Perhaps not surprisingly, humans have acquired a vocal tract unique in the animal world.

In mammals there are two basic patterns for the position of the larynx in the neck. The first is high

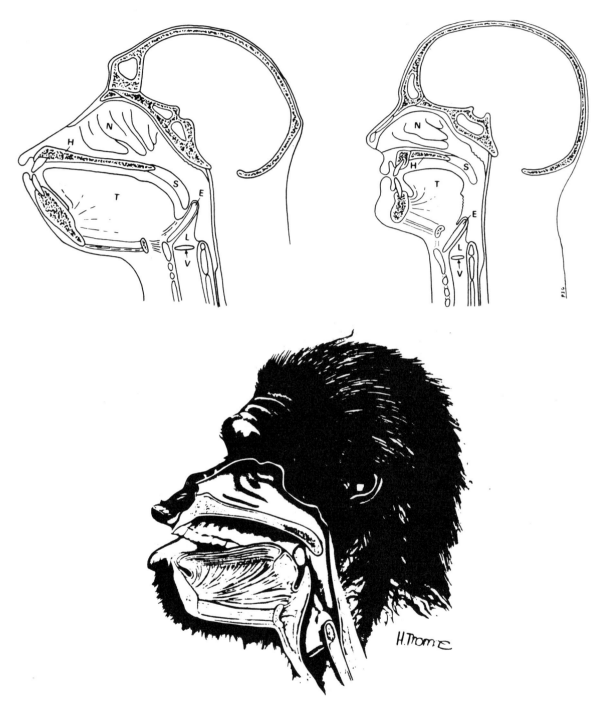

The vocal tract: Diagrams of the chimpanzee (above, left) and human (right) vocal tract: N = Nasal cavity; S = Soft palate; T = Tongue; P = Pharynx; L = Larynx; E = Epiglottis; V = Vocal fold. In the chimpanzee—as in all mammals—the larynx is high in the neck, enabling simultaneous breathing and swallowing. In mature

humans the larynx is lower in the next, making simultaneous breathing and swallowing impossible, but increasing the size of the pharynx and scope of vocal production. Below is sketch of the australopithecine vocal tract, which is like the chimpanzee's. (Courtesy of J. Laitman, Patrick Gannon, Hugh Thomas.)

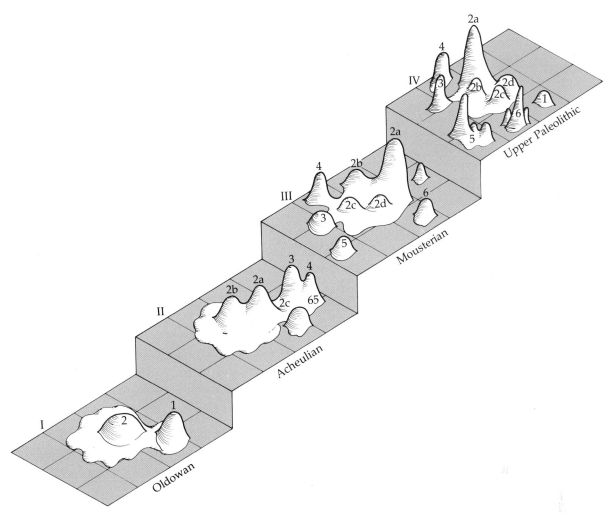

Sharpening the mind: sharpening the tongue: With the passage of time and the emergence of new species along the *Homo* lineage, stone-tool making became even more systematic and orderly. Peaks in the diagram represent identifiable artifact modes, with tall, narrow peaks implying highly standardized products. The increased orderliness in stone-tool manufacture must, argues archeologist Glynn Isaac, reflect an increasingly ordered set of cognitive processes that eventually involved spoken language. (I) e.g. Oldowan: 1 = Core-choppers; 2 = Casual scrapers. (II) e.g. Acheulian (Olorgesailie): 2a = Scrapers; 2b = Nosed scrapers; 2c = Large scrapers; 3 = Handaxes; 4 = Cleavers; 5 = Picks; 6 = Discoids. (III) e.g. Mousterian: 2a = Racloir; 2b = Grattoir; 2c = r. convergent; 3 = Percoir; 4 = Point; 5 = Burin; 6 = Biface. (IV) Upper Paleolithic: 2a = Grattoir; 2b = Nosed scraper; 2c = Raclette etc; 3 = Percoir; 4 = Point; 5 = Burins; 6 = Backed blades etc. (Courtesy of Glynn Isaac.)

up, which allows the animal simultaneously to swallow (food or liquid) and breath. The second pattern is with the larynx low in the neck, in which position the air passage has to be temporarily closed during swallowing, otherwise solids or liquids will block it and cause choking. Adult humans have the second pattern, while all other mammals, and infant humans, have the first.

As far as language ability is concerned, the low position of the larynx greatly enlarges the space above it: 'consequently, sounds emitted from the larynx can be modified to a greater degree than is possible for newborns and any nonhuman mammal', explains Laitman. Nonhuman mammals are limited to modifying laryngeal sounds by altering the shape of the oral cavity and the lips. Human newborns maintain the basic mammalian pattern until about 1.5 to 2 years, at which point the larynx begins to migrate lower in the neck, achieving the adult configuration by about 14 years.

Laitman and his colleagues discovered that the position of the larynx is reflected in the shape of the bottom of the skull, the basicranium: in humans it is arched, whereas in other mammals, and in human infants, it is much flatter. By looking at this feature in the fossil record it should therefore be possible to glean something about the verbal skills of extinct hominid species.

What does the fossil record say? 'In sum,' says Laitman, 'we find that the australopithecines probably had vocal tracts much like those of living monkeys or apes....The high position of their larynges would have made it impossible for them to produce some of the universal vowel sounds found in human speech.' Unfortunately, the fossil record for *Homo habilis* is poor as far as indications of the basicranium is concerned. But Laitman and his colleagues find that in its putative evolutionary successor, *Homo erectus*, 'the larynx...may have begun to descend into the neck, increasing the area available to modify laryngeal sounds'. The position of the larynx appears to be equivalent to that of an 8-year-old human. Only with the origin of archaic *Homo sapiens*, some 300,000 years ago, does the fully modern pattern appear, indicating at least the mechanical potential for the full range of sounds produced by people today.

So, indications from both lines of enquiry with fossil remains point to a rather gradual acquisition of language capabilities, although Holloway's interpretation would take it further back than is suggested by the basicranium evidence.

It should be remembered that higher primates are able to produce quite a wide range of sounds, which they use to subtle effect. For instance, when juvenile monkeys are threatened by an older opponent, they scream, which usually brings help. But the scream is subtly different, depending on the intensity of the threat and the dominance rank and kinship of the aggressor. And experiments with tape-recorded screams show that mothers respond to the scream according to the indicated danger. In addition, some higher primates have different alarm calls for different predators, leopard, snake, and so on: you can't say the different calls are 'words' but they do seem to be labels.

In thinking about the acquisition of spoken language by hominids, one therefore has to imagine the buildup of a greater and greater range of primate sounds, and their eventual conjunction as words. For some researchers, however, the structured use of words—syntax—which characterizes human speech is so different from primate vocalization that it is seen as disjunct. In other words, these researchers argue that human language is not on a continuum with primate vocalization.

We move now from hominid remains to the remains of things they made, which, says Holloway, is 'the only true "evidence" for early human cognitive behavior'. First, we will consider stone artifacts, objects made in the realm of economic activity; secondly, we will see what there is to learn from the more abstract realm of, for want of a better word, 'art'.

Some anthropologists have argued that the pattern of tool manufacture and language production—essentially, a series of individual steps—implies a common cognitive basis. In which case, following the trajectory of the complexity of stone-tool technology through time will also indicate something about the change in language capabilities.

Thomas Wynn, of the University of Colorado, has used psychological theory to examine the validity of this argument. 'It is true,' he says, 'that language and tool making are sequential behaviors, but the relationship is more likely to be one of analogy rather than homology.' In other words, there is a superficial similarity only, and their cognitive underpinnings are quite separate. You cannot look at the complexity of a tool assemblage on one hand and learn anything *directly* about language abilities on the other.

Glynn Isaac also looked for indications of language

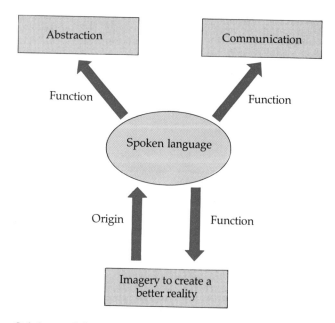

Origins and function of language: Although communication is clearly an important function of spoken language, its origins (and continued functions) probably centered on creating a better image of our ancestors' social and material worlds.

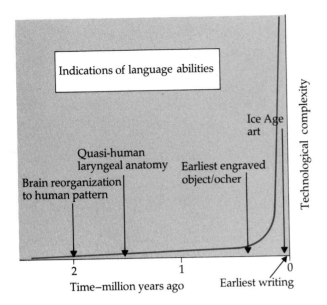

Indications of language abilities

Ice Age art

Quasi-human laryngeal anatomy

Brain reorganization to human pattern

Earliest engraved object/ocher

Technological complexity

Time—million years ago

2   1   0

Earliest writing

Cognitive indicators: The archeological record is restricted to material indications of language abilities, which, apart from writing, are necessarily indirect. Inferences from complexity in tool technologies can be added to fossil evidence of brain and vocal tract reorganization to give a basic indication of cognitive capabilities that might underlie language. Few observers would doubt that by the advent of wall and portable art, spoken language was already well developed.

function in ancient tool technologies, but with a different approach. He argued that you might be able to read in the complexity of a tool assemblage something about social complexity, not cognitive complexity relating to mechanical or verbal processes. Beyond a certain degree of social complexity there is an arbitrary imposition of standards and patterns. This is to some extent an abstract exercise, which would be impossible in the complete absence of language.

As we saw in units 24 and 25, the trajectory of technological change through hominid history falls into two phases: an incredibly slow phase leading from the earliest artifacts some 2.5 million years ago to about 200,000 years ago, after which an ever accelerating phase ensues.

What lessons do we learn from this basic archeological evidence, in relation to origins of language? Writ on the large scale, it seems reasonable to infer that a language complex enough to conjure the abstract elements of social rules, myths, and ritual is a rather late development in hominid history, beginning only with archaic *Homo sapiens*, and becoming fully expressed only with anatomically modern humans. If one adds the economic and

social organization necessary in hunting and gathering activities, which ultimately would involve the need for efficient verbal communication, then the archeological record shows that this follows the same pattern. Only in the later stages of hominid history does it take on a degree of sophistication the would seem to demand language skills.

We now turn from products of the hominid mind that concerned economic life to those of a more abstract character. Painting or engraving an image of, say, a bison, does not necessarily imply anything mystical about the motives in the mind of the artist. It is, nevertheless, an abstraction of the real world into a different form, a process that demands highly refined cognitive skills. But, as we saw in unit 26, the art of the ice age was not simply a series of simple abstractions of images to be seen in the real world. It was a highly selective abstraction, and, whether it represented hunting magic or an encapsulation of social structure, it therefore speaks of a world created by introspective consciousness and complex language. It was, in fact, a world like ours, just technologically more primitive.

If artistic expression can instruct about the possession of complex language, the question is, how far back in prehistory did it stretch? The answer is: not very far. The oldest abstract artifact known so far is a 300,000 year old ox rib from the site of Pech de l'Aze in France on which is carved a series of connected, festooned double arcs. This type of pattern is common in the Upper Paleolithic, from 40,000 years on, but between these two dates there is virtually nothing comparable. Of equal age to the engraved rib is the presence of red ocher within a coastal springtime shelter that has been excavated in southern France, at Terra Amata. Was this colored pigment used for body decoration, an activity that must be one of the most vivid yet least archeologically visible expressions of the transcendental human spirit? We will never know.

From Mousterian times, 150,000 to 40,000 years ago, there is a growing number of examples of engraved bone and ivory that indicate an awareness of an ability to deal with the abstract, as does the occurrence of obviously ritualistic burials. It is not until a little more than 30,000 years ago, however, that artistic expression really begins to blossom, slowly at first. One of the earliest sites is Vogelherd in Germany, from which were recovered several small, exquisitely carved animal figures, made from mammoth ivory. The most famous of the objects is the Vogelherd horse, dated at about 32,000 years.

Beyond Vogelherd, however, there is very little. Two pendants—one from reindeer bone, the other from a fox tooth—at the 35,000-year-old Neanderthal site of La Quina in France, and an antelope shoulder blade etched with geometric pattern from another French site, La Ferrassie. Elsewhere in Europe, bones and elephant teeth with distinct zigzag markings have been discovered, carved by Neanderthals at least 50,000 years ago.

Bearing in mind the probable imperfections in the archeological record—in Europe but more particularly in Africa—the inference to be drawn from artistic, abstract expression is that something important happened in the cultural milieu of hominids late in their history. The late British anthropologist Kenneth Oakley was one of the first to suggest, in 1951, that this 'something important' was best explained by a quantum jump in the evolution of language.

We know that language did evolve, perhaps with a trajectory that mirrored the complexity of material culture. The question now is: what was the evolutionary cause for its emergence?

The most obvious answer is that language evolved in the context of what it is so obviously proficient at: communication. For a long time this indeed was the line of argument pursued by anthropologists. The shift from the essentially individualistic subsistence activities of higher primates to the complex, cooperative venture of hunting and gathering surely demanded proficient communication. A popular part of the answer of how it came about included the notion that a first stage would have been a gesture language: gesturing, remember, is something humans do a lot, especially when lost for words.

In recent years, however, a shift of explanatory emphasis has occurred, paralleling the shift in explanation for the evolution of intelligence. From the practical world of communication, explanation of language origins now looks more to the inner world of thought and image making.

'The role of language in communication first evolved as a side effect of its basic role in the construction of reality', argues Harry Jerison. 'We can think of language as being an expression of another neural contribution to the construction of mental imagery....We need language more to tell stories than to direct actions.' As we saw in unit 27, anthropologists are beginning to see the importance of social interaction as the engine of the evolution of hominid intelligence. Consciousness and language go hand in hand with that.

Most of the expansion of hominid brain size occurred before material and abstract expressions of culture become really vibrant. This incremental expansion might be taken to imply an incremental buildup of consciousness and language in our ancestors, rather than a final, sudden bound, as we seem to see in the Upper Paleolithic. But there are many examples in biology of dramatic emergent effects as thresholds are passed: the origin of complex language and introspective consciousness might fit into this category.

Key questions:

• How important are the different lines of fossil evidence in revealing past language capabilities?
• How would one test the idea that conformity of stone-tool production implies the impositions of social rules, and therefore the existence of language?
• What kind of artistic expression is the most persuasive of the existence of language?
• If human language is discontinuous with primate vocalizations and communications, how might it have arisen?

Key references:

R.L. Holloway, 'Human paleontological evidence relevant to language behavior', *Human Neurobiology*, vol 2, pp 105–114 (1983).
G.L. Isaac, 'Stages of cultural elaboration in the Pleistocene', in *Origins and evolution of language and speech*, New York Academy of Sciences, 1976.
H.J. Jerison, 'Paleoneurology and the evolution of mind', *Scientific American*, January (1976).
A. Marshack, 'Implications of the paleolithic symbolic evidence for the origin of language', *American Scientist*, vol 64, pp 136–145 (1976).
New York Academy of Sciences, 'Origins and evolution of language and speech', *Annals*, vol 280 (1976).
R. White, 'Thoughts on social relationships and language in hominid evolution', *Journal of Social and Personal Relationships*, vol 2, pp 95–115 (1985).

29 / New worlds

There have been two major population dispersals in human history. The first was when *Homo erectus* moved out of Africa and into the rest of the Old World about 1 million years ago. The second was about 100,000 years ago, when recently evolved anatomically modern humans, *Homo sapiens sapiens* made the same journey. One difference between the two dispersals, however, was that *Homo sapiens* very quickly covered more territory than *Homo erectus*, not only reaching the frigid northern reaches of Eurasia, but also setting foot on the island continents of Australia and the Americas. Archeologists and anthropologists are still debating the dates at which these New Worlds were first occupied, but most would support something between 40,000 and 50,000 years ago for Australia, and about 12,000 years ago for the Americas.

There has often been a tendency to contemplate aspects of human history in isolation from that of other groups of animals. There must of course be some respects in which the path of human history has been determined solely by the rather special behavioral repertoire displayed by the genus *Homo*. Equally, however, the human lineage on occasions must have responded to ecological changes in ways parallel to that in other animals.

For example, Alan Turner of the Transvaal Museum in Pretoria argues that the initial dispersal from Africa and the later migration to North America can be viewed as territorial expansions in concert with other large predators. Rather than answering some inward spirit's urge for new lands, our ancestors were simply tracking their subsistence potential through new prey populations, as were other predators. However, one can only speculate on precisely what motivated the first Australian colonists to strike out in small boats for a land unseen.

The origin of the first Americans has been a matter of scientific debate for centuries, with Thomas Jefferson, for instance, speculating on an Asiatic linkage based on linguistic and archeological evidence. The source, Asia, and the route, across the Bering Strait that separates Alaska and Siberia, are agreed upon. But there is still dispute over the timing. One school of thought argues for something close to 12,000 years. There is no doubt that by 11,500 years the Americas had been peopled, as evidenced by the extensive archeological remains first of Clovis and then Folsom cultures. But were the Clovis people the first Americans? Not according to the second school of thought, which argues for a date in the region of 20,000 to 30,000 years. In recent years there has been a distinct shift in support of the first school.

Whenever they arrived, the first Americans found a land very different from the one we know today.

Migration routes to Australia and America: Eighteen thousand years before present was the apogee of the last glaciation (75,000–10,000 BP). Expanded glacial cover (white areas) lowered sea levels to expose shallow continental shelf (shaded areas over current coastlines). Although glaciation was less than shown 40,000 years ago, the Timor Straits were still considerably narrowed, facilitating the migration into Australia (and later into Tasmania). The reduced glaciation some 20,000–30,000 years ago might also have left an ice-free corridor linking North America and Siberia.

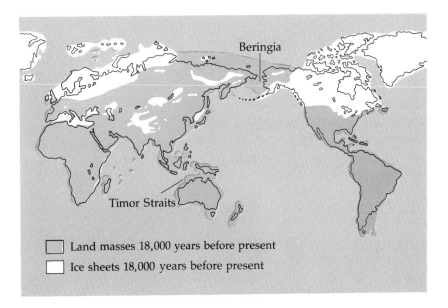

Beringia

Timor Straits

☐ Land masses 18,000 years before present
☐ Ice sheets 18,000 years before present

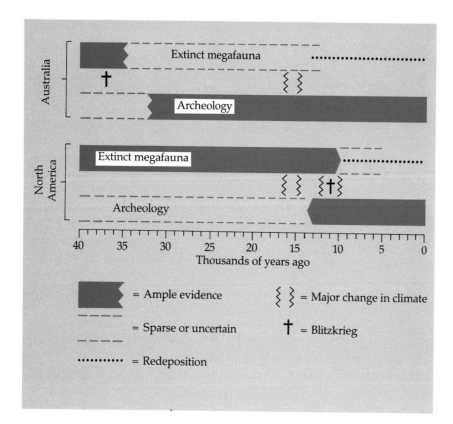

= Ample evidence   { { = Major change in climate

= Sparse or uncertain   † = Blitzkrieg

•••••••• = Redeposition

Indications of overkill?: In both the Americas and Australia the advent of human occupation (as indicated by archeology) closely coincides with a decline in large animal species (megafauna). Some scholars believe that this indicates that megafauna extinction was the result of overkill (blitzkrieg). (Courtesy of P. Martin.)

Between 75,000 and 10,000 years ago the earth was in the pulsating grip of an Ice Age, its frigid grasp being tightest at 65,000 and 18,000 years ago. Throughout this time North America was mantled with ice. The Laurentide ice sheet, 3 kilometers thick in places, buried much of Canada and the northern United States from the Atlantic coast to just east of the Rockies. The Cordilleran ice sheet ran ribbonlike up the Pacific coast from Washington State towards Alaska, submerging all but the highest peaks of the Rockies and the mountains of western Canada.

For all but the glacial maxima, an ice-free corridor appears to have linked southern North America with the ice-free regions of Alaska and Canada's Yukon and northwest territories, providing a potential migration route for people coming from Siberia. Such people would have made the intercontinental crossing dry shod, because the Beringia land bridge, which links Siberia with Alaska, was exposed for much of the time as the result of a drop in sea level, measuring as much as 100 meters at the glacial maxima, the water being locked up in the greatly expanded polar ice caps.

In principle, therefore, colonists from Eurasia could have made their way into the Americas at any time between 75,000 and 12,000 years ago. Some scholars argue, however, that although the ice-free corridor may not have represented a physical wall in the path of migrants, it might well have been so inhospitable as to be virtually biologically sterile. In other words, the ice-free corridor may nevertheless have been an invisible barrier. The time range for possible migration can probably be narrowed somewhat, because, according to archeological evidence so far available, Siberia remained uninhabited until between 30,000 and 20,000 years ago.

The Americas of the Ice Age were very different from today's world. They teemed with large mammal species, including mammoth, mastodon, giant ground sloth, steppe bison, elk, yak, lion—75 species in all, many of which were immigrants from Eurasia. Huge freshwater lakes ponded in the Great Basin. And the great equatorial forests of Central and South America survived in sheltered 'refuges', having largely been replaced by open grassland and woodland.

Clovis people, who made a characteristic 'fluted' projectile point, an American invention, lived in the narrow archeological window between 11,500 and 11,000 years ago. They were replaced by Folsom people, who produced smaller, more finely crafted projectile points. But the Clovis and Folsom worlds were very different places. Clovis people hunted mammoth and mastodon. By Folsom times, none remained. Gone too were the great majority of large mammals, 75 species eventually going extinct or becoming restricted to South America.

One of the great debates over the peopling of the Americas has centered on this rapid extinction. Some authorities, Paul Martin of the University of Arizona being most prominent among them, argue that the animals had been wiped out by a wave of Clovis and then Folsom hunters, advancing north to south for a millennium. Others, with Ernest Lundelius of the University of Texas most prominent, point to the dramatic climatic shift at the end of the ice age as the culprit.

Invasion of new lands by humans has been known to cause significant extinctions in relatively recent history. And climate change can certainly drive species to extinction, particularly a change as dramatic as that marked by the Pleistocene/Holocene transition. So, while both explanations are plausible, neither has been demonstrated beyond reasonable doubt in the case of the Ice Age mammals of the Americas. (It has to be noted, however, that for only seven of these species is there good fossil evidence, indicating that extinction occurred in the Clovis 'window'. It remains possible that, for some of the rest, extinction occurred in pre-Clovis times.)

As Donald Grayson of the University of Washington has pointed out, the image of Clovis people as mammoth hunters may have been overemphasized through bias of the archeological record. 'Most Clovis sites have been uncovered following

Clovis and after: Although skeletal remains are few, the Clovis people left their trade mark—the Clovis point (far left)—widely over North America. The Clovis point, which usually measured about 7 cm in length, was apparently inserted into the split end of a spear shaft, and bound in place by hide. Cultures that followed in close succession after Clovis were (second-left to right) Folsom, Scottsbluff, and Hell Gap.

The time of Clovis: Clovis sites are scattered over much of North America (specifically the United States, as most of Canada was under ice at the time). As this diagram shows, dating of the sites lies in a tight range between 11,500 and just less than 11,000 years ago. Folsom sites follow close on behind, but again confined to North America

12,000	11,000	10,000	Time scale in radiocarbon years, BP	
			Walker Road Moose Creek Owl Ridge Dry Ceek	Alaska
			Agate Basin pre-Folsom Mill iron U.P. Mammoth Colby Dent Clovis Lange/Ferguson Domebo Lehner Murray Springs	Clovis sites
			Lindenmeier Hell Gap Agate Basin Clovis Folsom Hanson Bonfire	Folsom sites

the initial discovery of large bones', he says. 'As a result, if Clovis people in the west spent most of their time hunting mice and gathering berries, we would probably not know it.' In any case, none of the discoveries of Clovis material east of the Mississippi is associated with signs of big-game hunting.

But the most tantalizing question is, who, if anybody, preceded the Clovis people in the Americas? Richard Morlan, of the Canadian Museum of Civilization, Ottawa, has surveyed the 50 or more sites 'south of the ice' with ages putatively older than 11,500 years, and concludes that 'all earlier sites are controversial to some degree'.

It should be noted first that some of the more

famous 'old' sites have recently lost their claims at predating Clovis. Calico Hills, in California, which its proponents claim yields stone artifacts dating between 100,000 and 200,000 years, is no longer taken seriously by most authorities. Del Mar Man, a collection of skulls once dated at 70,000 years have recently been redated at about 8000 years. And the famous bone deflesher from Old Crow in the Yukon territories, found in 1966 and dated at 27,000 years, was redated in 1987 at just 1400 years. Nevertheless, Morlan believes that another Yukon site, Bluefish Caves, may prove to be in the range of 25,000 years.

The serious pre-Clovis contenders south of the ice are mostly in South America. They are: Los

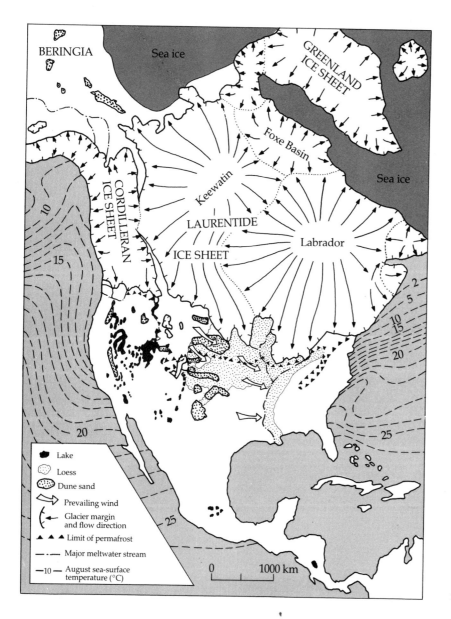

In the grip of the ice: At the peak of the last glaciation, about 18,000 years ago, much of North America was covered by thick ice sheets: to the west was the Cordilleran Ice Sheet, in center and east, the Laurentide Ice Sheet. There is still dispute as to whether an ice-free corridor existed throughout the period, or was temporarily closed. (Courtesy of Stephen C. Porter.)

Toldos Cave in the Argentine Patagonia, dated at 12,600 years; the site of Tagua-Tagua in central Chile, dated at 11,380; also in central Chile, the site of Monte Verde, dated at 13,000; and Taima-Taima in northwestern Venezuela, dated at 13,000. The most important site in North America, and probably the strongest contender of all, is the Meadowcroft cave shelter near Pittsburgh, a site that is said to have been occupied repeatedly since 19,600 years ago.

Skeptics of Meadowcroft's putative 19,600 year date point out the possibility of contamination of the site's material with carbon from nearby coal deposits, which would disrupt the radiocarbon dating used at the site. James Adovasio of the University of Pittsburgh, the site's principal investigator, counters by noting that the dates run from oldest to the youngest in the deposits from the bottom to the top in the site, just as they should if they were uncontaminated. The issue remains to be resolved.

But, asks Vance Haynes of the University of Arizona, if Meadowcroft is an authentic pre-Clovis site, why are there so few other potential sites from this period? And why did it take so long for the population to expand? There's no doubt that, once the Clovis culture arrived, population expanded at an enormous rate: rising from a few hundred people to over a million in 300 years, a growth rate of about 3 per cent per annum.

When Columbus arrived in the Americas, 1000 different languages were spoken among the native Indian peoples. Stanford University linguist Joseph Greenberg has analyzed the 600 languages that survive, and traces them back to just three: Amerind, the most widespread and diversified; Na-Dene, less widespread and diversified than Amerind, but more so than the third group; Aleut–Eskimo. It is possible, says Greenberg, that these three linguistic groups represent three separate migrations, with the Amerind group being the first arrivals. Several molecular biology laboratories are working on mitochondrial DNA analysis, which might help answer questions about the timing and number of the first Americans.

By contrast with their American cousins, the first Australians had to make a water journey to their New World. Even with sea levels at their lowest during glacial maxima, the journey from Sunda Land (the combined landmass of Southeast Asia and much of Indonesia) to the Sahul landmass (Australia, Tasmania, and New Guinea) would have required eight sea voyages, the last being 87 kilometers. So far, there is no archeological evidence from Australian sites of vessels that could have made such a journey. However, what were coastal sites during the ice age are mostly now submerged beneath the sea.

Hominids have been in southeast Asia for almost a million years, but the first evidence of occupation in the Sahul is just 40,000 old, an archeological site on the northeast coast of New Guinea. Within Australia itself, archeological and fossil evidence comes mainly from the west and the south, far away from the putative landfall sites. The oldest archeological site is at Lake Mungo in New South Wales, and dates to as much as 45,000 years. The oldest human fossils come from close by, at Willandra Lakes, and seems to be at least 30,000 years old.

Human fossils—mainly crania—from Australia have often been said to fall into two groups: anatomically gracile individuals, like those at Lake Mungo and Keilor, in southern Victoria; and more robust people, such as those at Kow Swamp and Cohuna, also in Victoria. This putative division

Australian evidence: Major archeological and fossil sites (with dates where known) are shown here. Most scholars put the colonization of Australia at some time between 40,000 and 50,000 years ago. The shaded areas show continental shelf exposed under peak glaciation.

prompted the notion of two separate migrations, the gracile colonist being said to have come from China, and the robust colonists from Indonesia. Interbreeding would have blurred the distinctions in later generations.

In fact, the division of the earliest fossils into gracile and robust is somewhat artificial, argues Phillip Habgood, of the University of Sydney. He, and a growing number of Australian scholars, suggest that the early colonists were more anatomically homogeneous, the variable morphology of the modern Aborigines being the result of 'genetic (and cultural) processes acting upon a small founding population'.

As indicated in unit 22, some scholars, such as Milford Wolpoff, of the University of Michigan, and Alan Thorne, of the Australian National University, Canberra, argue for deep-rooted local continuity for modern humans. In this case, Australian ancestors would be derived from 200,000-year-old populations in central Java, to which the archaic sapiens fossils from Ngandong belonged. However, evidence from mitochondrial DNA (unit 23) indicates a more recent African origin for Australian Aborigines. The same evidence, incidentally, implies that the island continent was colonized at least 15 times, not just once.

Key questions:

● How important are the earliest Siberian sites in setting limits on the first entry into North America?
● What archeological signature would be expected if Paul Martin's 'overkill hypothesis' were correct?
● How could the overkill hypothesis best be tested?
● Anatomical evidence for the colonization of Australia indicates two founder populations at most, while mitochondrial DNA evidence suggests 15: how is the disparity best explained?

Key references:

R. Carlisle (editor) *Americans before Columbus: perspectives on the archeology of the first Americans*, University of Pittsburgh Press, 1988.
Joseph H. Greenberg *et al.*, 'The settlement of the Americas', *Current Anthropology*, vol 27, pp 477–497 (1986).
Phillip Habgood, 'The origin of Australian Aborigines', in *Hominid evolution: past, present and future*, edited by Phillip Tobias, Alan R. Liss Inc., 1985.
The first Americas, series of articles in *Natural History* between November 1986 and January 1988.
Alan Turner, 'Hominids and fellow travelers', *South African Journal of Science*, vol 78, pp 231–237 (1982).

30 / The first villagers

The date of 12,000 years before present (BP) is usually given as the beginnings of what has been called the Agricultural (or Neolithic) Revolution. Prior to this date human populations subsisted by various forms of hunting and gathering. From 12,000 BP, however, a shift to plant and animal domestication occurred independently in several different parts of the world, first in the Fertile Crescent of the Near East, then in southern Asia, and last in various locations of the New World. The adoption of agriculture was extremely rapid as measured against the established timescale of human prehistory, and was accompanied by an escalating increase in population, rising from about 10 million at the outset of the Neolithic to 100 million some 4000 years ago.

The tremendous changes wrought during the Neolithic period can be seen as a prelude to the emergence of cities and city states; and of course to a further rise in population, to the 6 billion figure of today.

Until relatively recently, the Agricultural Revolution was viewed as a rather straightforward—if dramatic—transition. Responding to some kind of stimulus, hunter—gatherers, who were assumed to have lived in small nomadic bands of about 25 individuals, took to plant and animal domestication as a way of intensifying food production. As a result, the argument ran, people began living in large, settled communities whose social and political complexity far exceeded anything achieved earlier in history. In other words, sedentism and social complexity were explained as the *consequences* of the adoption of agriculture. And the Neolithic transition was characterized as a shift from the simple to the complex.

In the light of new archeological and ethnographic evidence, and a reassessment of some existing evidence, however, the Neolithic transition is now viewed rather differently. Most importantly, it is now clear that in many cases populations established sedentary communities and elaborated complex social systems *prior to* the advent of agriculture. Hunter—gatherers of the Late Pleistocene, it is now realized, were not necessarily living the simple, nomadic lifeway that anthropologists had imagined. Although there remains debate about what triggered the Neolithic transition, it is not unreasonable to view agriculture in some ways as a consequence, not the cause, of social complexity.

The traditional characterization of the Neolithic

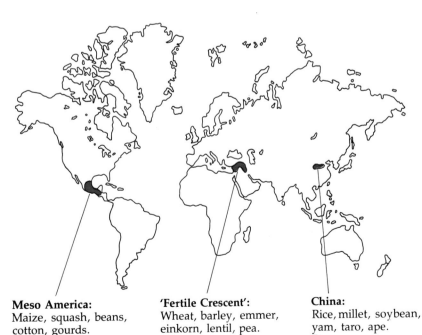

Major centers of agricultural innovation: Plant and animal domestication apparently occurred independently and at different times in many different parts of the world. There were, however, three major centers of origin, whose influence spread geographically, eventually coming to dominate local innovations.

Meso America:
Maize, squash, beans, cotton, gourds.
Llama, guinea-pig
[5000 years ago]

'Fertile Crescent':
Wheat, barley, emmer, einkorn, lentil, pea.
Goats, sheep, cattle
[10,000 years ago]

China:
Rice, millet, soybean, yam, taro, ape.
Pigs,
[7000 years ago]

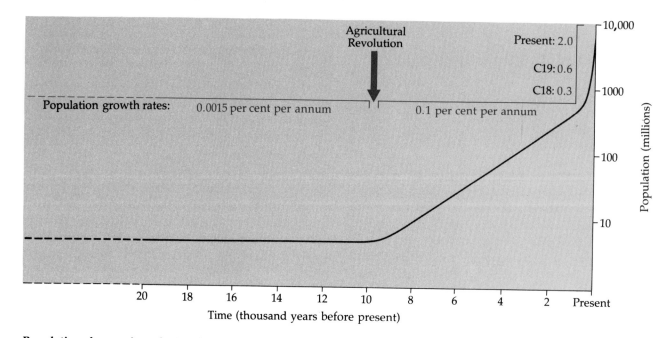

Population growth rates: 0.0015 per cent per annum   0.1 per cent per annum

Agricultural
Revolution

Present: 2.0

C19: 0.6

C18: 0.3

Population (millions)

Time (thousand years before present)

Population change since the Neolithic: There is no doubt that the beginnings of substantial population growth coincided with the origin of plant and animal domestication, igniting an explosion that continues today. But there is still dispute over whether population growth itself was a cause or a consequence of domestication.

transition as an Agricultural Revolution rested on two kinds of evidence: archeological and ethnographic, the former seen as indicating an explosive change in economic organization, the latter as revealing a shift from simple to complex social organization.

The phrase 'Agricultural Revolution' had been coined in the 1950s by the Australian prehistorian V. Gordon Childe. It had seemed an apt term for a number of reasons, not least of which was the limited amount of archeological data with which to sketch this crucial period in human history. The few major sites, such as the early farming and trading community of Jericho, with its impressive tower and wall, seemed to burst out of an archeological void with dramatic suddenness. Indeed, the remains of Catal Huyuk, which was occupied by farming people between 8500 BP and 7800 BP, has been described as an archeological supernova. Excavated in the 1960s, this Turkish town of about 30 acres boasted elaborate architecture and beautiful, symbolic wall paintings and carvings.

In the past decade and a half, however, further excavations in the Fertile Crescent have uncovered the remains of villages and towns, which collectively make clear that the adoption of agriculture was a much more gradual process than had been envisaged. Such sites include 'Ain Ghazal in Jordan, Gritille in Turkey, and Abu Huyrera in northern

Syria. In particular, there can now be seen a transition from settled communities based entirely on hunting and gathering, through a mixed economy of hunting and gathering combined with some domestication, to fully committed agriculture. In the closer focus offered by a more complete archeological record, the Neolithic transition can therefore be seen as a step by step introduction of domestication, not an overnight revolution.

One of the most informative sites is that of Abu Hureyra, which was occupied from 11,500 BP through to 7000 BP, with one major break from 10,100 to 9600 BP. Emergency excavation in 1974 showed that the first period of settlement, Abu Hureyra I, was a hunting and gathering community of between 50 and 300 individuals who exploited the rich steppe flora (including many wild cereals) and the annually migrating Persian gazelle. A year-round settlement of simple yet substantial single-family houses, Abu Hureyra I confounds the traditional view of hunter–gatherer existence: that of the small, nomadic band.

Perhaps because of over-exploitation of local resources, and an increasingly unfavorable climate, Abu Hureyra I was abandoned in 10,100 BP, to be reoccupied about half a millennium later, this time by people who included plant—but not animal—domestication in their economy. For a millennium, the people of Abu Hureyra continued to hunt gazelle as their sole source of meat, after which time they turned to the domestication of sheep and goats. The overall pattern, therefore, is 'a step by step introduction of domesticated plants and animals', explains Yale University's Andrew Moore, who led the 1974 excavation. 'This is a pattern I see across southwestern Asia.'

It really should have come as no surprise that Late Pleistocene hunter—gatherers led socially complex lives, as indications of this have been known from the archeological record for some time. Most notable among this evidence was the art of the European Ice Age (see unit 26). 'If one is looking for a single archeological reflection of sociocultural complexity, then presumably attention will continue to focus on the unique and impressive manifestations of Upper Paleolithic cave art from the Franco-Cantabrian region', notes Paul Mellars of Cambridge University, England. This period of wall and portable art began about 30,000 years ago, reached a climax 15,000 years ago, and ended 10,000 years ago, the termination of the ice age.

More tangible evidence of Late Pleistocene social and economic complexity comes from the Central Russian Plain, specifically a site near the town of Mezhirich, 1100 kilometers southwest of Moscow. Fifteen thousand years ago a settlement of some 50 people lived in a 'village' of at least five substantial dwellings, each constructed from mammoth bones. 'We are beginning to find evidence of semipermanent dwellings in the Central Russian Plain dating back to nearly 30,000 years ago', notes Olga Soffer of the University of Wisconsin.

With this and other evidence, it is perhaps surprising that until relatively recently Late Pleistocene humans were almost universally regarded as simple nomads, endlessly wandering from camp to camp in bands of no more than 25 individuals. One reason for this characterization was the very important and influential study during the 1960s of the !Kung San (Bushmen) of the Kalahari. Organized from Harvard University by anthropologists Richard Lee and Irven DeVore, the !Kung project examined in great detail the socioeconomic life of these people.

The project revealed that, in spite of living in a marginal environment, the !Kung were able to subsist on simple hunting and gathering, with the expenditure of just a few hours work each day. In addition, !Kung social life was characterized as an egalitarian, harmonious, sharing environment. The collective results of the Harvard project were

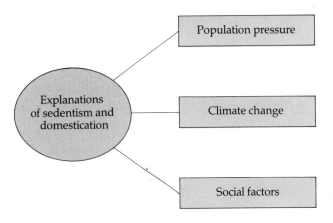

Hypotheses of agricultural origins: Population pressure and climate change have long vied as the most persuasive potential candidate for initiating sedentism and domestication. In more recent times, attention has turned to factors concerning internal social complexity.

presented at a landmark meeting, titled 'Man the Hunter', held at the University of Chicago in 1966. For several reasons—including the fact that no other ethnographic project had been so thoroughly and scientifically conducted—the Harvard project's portrayal of the !Kung became *the* image of the hunter—gatherer lifeway, both in the modern world and in prehistory. This, despite existing archeological and ethnographic evidence to the contrary.

For more than a decade the !Kung model of the hunter—gatherer lifeway dominated anthropological thought, but by the early 1980s its shortcomings gradually came to be exposed. This shift in perception was driven by new historical, archeological, and behavioral ecology evidence: the new evidence indicated, first, that there was a great deal more variability in the hunter—gatherer lifeway than had been allowed for in the !Kung model; and, second, that this variability included a degree of social and economic complexity that hitherto had been associated exclusively with agricultural societies.

'Many characteristics previously associated solely with farmers—sedentism, elaborate burial and substantial tombs, social inequality, occupational specialization, long-distance exchange, technological innovation, warfare—are to be found among many foraging societies', concluded anthropologists James Brown and T. Douglas price in 1984, in a classic reassessment of hunter—gatherers.

In other words, the Agricultural Revolution was recognized as being neither a revolution nor primarily focused on the adoption of agriculture. The

Neolithic transition involved increasing sedentism and social complexity, which was usually followed by the gradual adoption of plant and animal domestication. In some cases, however, plant domestication preceded sedentism, particularly in the New World. For instance, Kent Flannery of the University of Michigan has shown that the first plant domesticate in the New World, the bottle gourd, which was grown about 9000 years BP in the southern highlands of Mexico, preceded sedentism by at least 1000 years. Clearly, the Neolithic was a complex period, one that must have been influenced substantial by local as well as global factors.

At least as important as understanding the trajectory of the Neolithic transition, of course, is insight into its underlying cause. Because the transition occurred within a few thousand years independently in several different parts of the world, there has long been a desire to search for a global cause. Two factors have been candidates for this single, prime mover: population pressure and climate change.

There is no question that a dramatic rise in population numbers accompanied the Neolithic transition, but there remains the question of whether this relationship was one of cause or effect. Mark Cohen of the State University of New York, Platts-burgh, the principal proponent of the population pressure hypothesis, argues that it was causal, and adduces signs of nutritional stress in skeletal remains from the Late Paleolithic to support his case. However, many anthropologists argue that there are many examples of the adoption of sedentism and agriculture in the apparent absence high population numbers—such as in the southern highlands of Mexico. For these researchers, including Flannery, the population pressure hypothesis remains unconvincing.

The second major candidate—climatic change—looks persuasive, because the Neolithic transition coincides with the end of the Pleistocene glaciation. The shift from glacial to interglacial conditions would have driven extensive environmental restructuring, bringing plant and animal communities into areas where they did not previously exist. For instance, there seems little doubt that warmer, moister climes in the Levant 12,000 years ago encouraged the abundant growth of wild cereals on the steppe, so that initially foragers could collect them in great numbers, and subsequently domesticate them. Moore considers this to have been important in the early establishment of Abu Hureyra and other similar settlements.

However, evidence is lacking to argue that climate-driven floral change universally accompanied

Traditional view of agricultural revolution

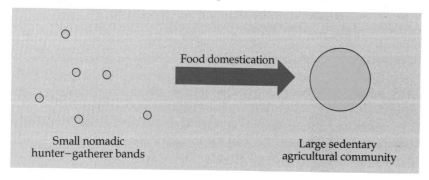

Small nomadic hunter–gatherer bands   Food domestication   Large sedentary agricultural community

Current view of agricultural revolution

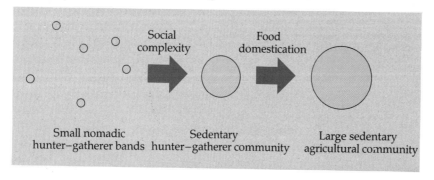

Small nomadic hunter–gatherer bands   Social complexity   Sedentary hunter–gatherer community   Food domestication   Large sedentary agricultural community

Origin seen as more complex: In the traditional view, sedentism and domestication developed together: it was a switch from small, nomadic, hunter–gatherer bands on one hand to large, sedentary, agricultural communities on the other. Recently, scholars have come to realize that the process probably included several steps, in which sedentism and domestication were separated. Intermediate between small nomadic bands and large, agricultural communities, therefore, were sedentary communities that subsisted on hunting and gathering.

sedentism. Moreover, there must have been periods earlier than the end of the Pleistocene that would have been conducive to intensification of food production. Modern *Homo sapiens* arose more than 100,000 years ago, so why did almost 90,000 years pass before intensification of food production become adopted? Perhaps it was a combination of population pressure and climate change? Or perhaps it was something else entirely.

For some scholars, that 'something else' is social complexity. Whereas population pressure and climate change were both 'external' factors—the first presenting a problem to be solved, the second an opportunity to be exploited—social complexity would be an 'internal' trigger for change.

Building on earlier ideas of Robert Braidwood, University of London anthropologist Barbara Bender argues that social complexity is a prerequisite to—not a product of—a sedentary agricultural system. The increasing social complexity, and the stratified social and economic order that goes along with it, make demands on food production that cannot be satisfied by the small, nomadic hunter—gatherer

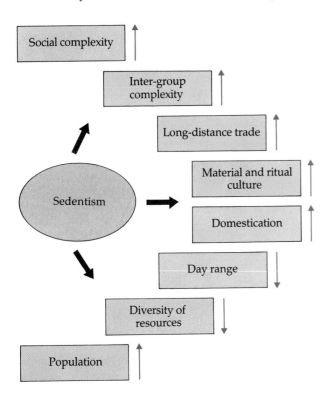

Consequences of sedentism: A shift from a nomadic to a sedentary way of life necessarily involved a series of potential social and material changes. Although these changes have often been associated exclusively with agricultural societies, it is now evident that sedentism *per se* will produce at least part of this pattern.

society, Bender and her supporters say. The response to this internal pressure is to intensify and formalize food production: in other words, the agrarian society. Bender is not arguing that this internal factor is the sole cause, merely that 'Technology and demography have been given too much importance in the explanation of agricultural origins, social structure too little'.

Although this social focus is gaining popularity among anthropologists, assessing its merits is very difficult: like a black box, you know it is important, but don't know how it works. Why, for instance, would social complexity have taken 90,000 years to manifest itself after the origin of anatomically modern humans? One possibility, of course, is that a subtle intellectual evolutionary change may have occurred rather recently in human history, one that does not manifest itself physically.

In fact, there was a biological change among modern humans between the end of the Pleistocene and the Holocene, but this was in bodily physique. Not only are post-Pleistocene humans smaller than their immediate ancestors, but the difference in size between males and females—sexual dimorphism—is significantly reduced. As Robert Foley of Cambridge University, England, has recently pointed out, this may have implications for how one views the Neolithic transition.

Inevitably, anthropologists' concept of hunter—gatherers is influenced by what is known of contemporary foragers. These people, whose numbers are rapidly disappearing and who live in the most marginal areas of the globe, generally include a large plant-food component in their diet (there are notable exceptions, of course) and exist in egalitarian communities. So, the Neolithic transition is usually seen as one of a change from this kind of subsistence economy to domestication.

However, the larger overall body size of Late Pleistocene people, and the greater sexual dimorphism in body size, might imply a different socioeconomic context. There may well have been more competition between males for access to females (see unit 11), and the males may have engaged in more big-game hunting and provisioning of their mates and offspring. 'In this context, what we think of as modern hunter—gathering is largely a post-Pleistocene phenomenon', says Foley. 'Rather than being an adaptation ancestral to food production, it is a parallel development....Both hunter—gatherer and agricultural systems developed as a response to resource depletion at the end of the Pleistocene

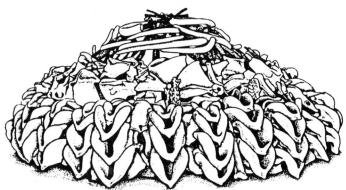

Mammoth-bone dwelling: This dwelling, which measured 5 meters in diameter, is one of five excavated at Mezhirich, in the Ukranian Republic of the USSR. Individually contructed with great technical and esthetic attention, these 15,000-year-old dwellings formed a community that was surely more socially complex than is usually envisaged for pre-agricultural hunter–gatherer peoples. (Courtesy of M.I. Gladkih, N.L. Kornietz, and O. Soffer/*Scientific American,* Novemer 1984, all rights reserved.)

from the rather different socioecology of Late Pleistocene anatomically modern humans.'

Clearly, anthropologists' picture of the Neolithic transition is far from complete. It is fair to say, however, that the search for a single, prime mover is much less popular than it was. 'No single factor is responsible for the rise of cultural complexity', concluded Brown and Price. 'Increased complexity appears in too many diverse and historically unconnected places to be a result of a single factor. . . . It may be sufficient for the moment simply to be aware that things are not what they have seemed to be.'

Key references:

Kent Flannery (editor), *Guila Naquitz,* Academic Press, 1986. (Papers on the research project in the Valley of Oaxaca, Mexico.)

Robert Foley, 'Hominids, humans and hunter–gatherers: an evolutionary perspective', in *Hunters and gathers: history, evolution and social change,* edited by T. Ingold, D. Riches, and J. Woodburn, Oxford University Press, 1988.

Mikhail I. Gladkih *et al.,* 'Mammoth-bone dwellings on the Russian Plain', *Scientific American,* November 1984.

Richard Lee and Irven DeVore, *Man the hunter,* Aldine, 1968. (Papers from the famous 1966 University of Chicago conference, including work on the !Kung.)

Anthony J. Legg and Peter A. Rowley-Conwy, 'Gazelle killing in stone age Syria', *Scientific American,* August 1987. (The most recent account of Abu Hureyra.)

Roger Lewin, 'A revolution of ideas in agricultural origins', and 'New views emerge on hunters and gatherers', *Science,* vol 240, pp 984–986 and 1146–1148 (1988).

'Prehistoric Pioneers', special issue of *Expedition,* The University Museum Magazine of Archeology/Anthropology, University of Pennsylvania, vol 28, number 2 (1986).

T. Douglas Price and James A. Brown (editors), *Prehistoric hunter gatherers,* Academic Press, 1984. (A series of papers on sociocultural complexity in foragers of the Old and New Worlds.)

A.H. Simmons *et al.,* ''Ain Ghazal: a major Neolithic settlement in Central Jordan', *Science,* vol 240, pp 35–39 (1988).

Glossary

Anthropoid: the informal name for the suborder of primates that includes monkeys, apes and humans.

Analogy: similarities among organisms based on convergent evolution (contrast with homology).

Arboreal: tree-living.

Autapomorphy: a derived character not shared by other species.

Bipedalism: upright walking on the two hind legs, as in humans.

Brachiation: mode of locomotion through trees, using the arms for hanging and swinging, as in gibbons.

Carnivore: a meat-eating animal.

Clade: a group of species that contains the common ancestor of the group and all of its descendants.

Cranium: the complete skull (brain case, face and palate, and lower jaw) (see postcranium).

Cusps: the conical projections on the surface of teeth.

Derived character: a character acquired by some members of an evolutionary group, and therefore serves to unite them in a taxonomic sense and distinguish them from other species in the group (contrast with primitive character).

Endocast: the impression of the inner surface of the brain case; can be natural or experimentally made.

Fauna: the animal component of an ecosystem.

Flora: the plant component of an ecosystem.

Folivore: a leaf-eating animal.

Founder effect: the genetic consequences in a population that is established from a small number of individuals, perhaps one male and one female.

Frugivore: a fruit-eating animal.

Genotype: the genetic makeup of an organism (contrast with phenotype).

Gradualism: mode of evolution that involves the steady accumulation of small changes (contrast with punctuationalism).

Hominid: the informal name for the Hominidae, the human family, as currently classified.

Hominoid: humans and apes.

Homoplasy: a character shared between species as a result of convergent evolution.

Kin selection: the genetic consequences of the behavior of one individual that enhances the reproductive success of its relatives.

Lineage: an evolutionary line linked by common ancastry.

Lower Paleolithic: human cultural period, beginning with the first appearance of stone tools and ending about 150,000 years ago, usually referring to Europe; the African equivalent is Early Stone Age.

Macroevolution: large changes in the evolutionary record, such as trends in groups and the emergence of new groups.

Mandible: lower jaw.

Maxilla: upper jaw.

Microevolution: small evolutionary changes, often within populations or a single species.

Middle Paleolithic: part of the Old Stone Age, from about 150,000 years ago until about 50,000 years ago; the term is usually applied to European cultures, the equivalent African culture being the Middle Stone Age.

Monogamy: social system in which mates have reproductive access to only one partner.

Morphology: the physical form of an organism.

Neolithic: the New Stone Age, usually associated with the beginnings of agriculture.

Omnivore: a species that includes a range of food types in its diet.

Paleolithic: the Old Stone Age, starting with the first appearance of stone tools in the archeological record, and ending with the beginnings of agriculture (see Lower Paleolithic, Middle Paleolithic, and Upper Paleolithic).

Phenotype: the physical characteristics of a species (see genotype.)

Phyletic change: the evolution of a new species through the gradual change of an existing species, resulting in no increase in species diversity.

Phylogeny: the evolutionary history—family tree—of organisms.

Polygyny: social structure in which one male has reproductive access to more than one famale.

Postcranium: all that part of the skeleton that excludes the cranium.

Primitive character: a character that was present in a common ancestor of a group and is therefore shared by all members of that group (contrast with derived character).

Prognathous: having the jaws protruding in front of the line of the upper face.

Punctuationalism: mode of evolution in which changes are concentrated into geologically brief periods (contrast with gradualism).

Terrestrial: ground-living.

Reproductive success: a measure of an individual's genetic representation in the next generation, and includes the number of offspring produced and their survival.

Sexual dimorphism: the state in which some aspect of a species' anatomy consistently differs in size or form between males and females.

Sociobiology: an evolutionary approach to the explanation of behavior, with particular emphasis on kin selection.

Socioecology: the study of the interaction between a species' behavior and its environment, which also includes other species.

Speciation: the evolution of a new species through the splitting of an existing lineage, thus increasing species diversity.

Stone Age: the earliest period of human culture, from about 2.5 million years ago until the first use of metal, about 5000 years ago; divided into Old Stone Age and New Stone Age.

Symplesiomorphy: a shared, primitive character.

Synapomorphy: a shared, derive character.

Systematics: the science of classification, as taxonomy.

Taxonomy: classification of organisms according to evolutionary relationship.

Upper Paleolithic: the cultural period beginning about 50,000 years ago and ending 10,000 years ago, usually with reference to Europe; the equivalent in Africa is the Later Stone Age.

Index